D0913137

BLACK HOLES

BRIAN COX
JEFF FORSHAW

BLACK HOLES

THE KEY TO UNDERSTANDING THE UNIVERSE

MARINER BOOKS
New York   Boston

Images © HarperCollins*Publishers* 2022 unless otherwise indicated in
Picture Credits. Illustrations by Martin Brown and Jack Jewell.

HarperCollins books may be purchased for educational, business,
or sales promotional use. For information, please email the Special
Markets Department at SPsales@harpercollins.com.

Originally published in the United Kingdom in 2022 by William
Collins, an imprint of HarperCollins Publishers.

FIRST U.S. EDITION

Library of Congress Cataloging-in-Publication Data has been applied
for.

ISBN 978-0-06-293669-1

23 24 25 26 27 LBC 5 4 3 2 1

For Jeff's mum, Sylvia

CONTENTS

1

A BRIEF HISTORY OF BLACK HOLES

'A knowledge of the existence of something we cannot penetrate, of the manifestations of the profoundest reason and the most radiant beauty – it is this knowledge and this emotion that constitute the truly religious attitude; in this sense, and in this alone, I am a deeply religious man.'

Albert Einstein

At the heart of the Milky Way, there is a distortion in the fabric of the Universe caused by something 4 million times more massive than our Sun. Space and time are so warped in its vicinity that light rays are trapped if they venture closer than 12 million kilometres. The region of no return is bounded by an event horizon, so named because the Universe outside is forever isolated from anything that happens within. Or so we used to think when the name was coined. We have named it Sagittarius A* and it is a supermassive black hole.*

Black holes lie where the most massive stars used to shine, at the centres of galaxies and at the edge of our current understanding. They are naturally occurring objects, inevitable creations of gravity if too much matter collapses into a small enough space. And yet, although our laws of Nature predict them, they fail to

---

* Sagittarius A* is pronounced 'Sagittarius A-star'.

fully describe them. Physicists spend their careers looking for problems, conducting experiments in search of anything that cannot be explained by the known laws. The wonderful thing about the increasing number of black holes we have discovered dotted across the sky is that each one is an experiment conducted by Nature that we cannot explain. This means we are missing something deep.

The modern study of black holes begins with Einstein's General Theory of Relativity, published in 1915. This century-old theory of gravity leads to two startling predictions: 'First, that the fate of massive stars is to collapse behind an event horizon to form a "black hole" which will contain a singularity; and secondly, that there is a singularity in our past which constitutes, in some sense, a beginning to the universe.' This remarkable sentence appears on the first page of a seminal textbook on general relativity, *The Large Scale Structure of Space-Time*, written in 1973 by Stephen Hawking and George Ellis.[1] It introduces evocative terms – black hole, singularity, event horizon – which have become part of popular culture. It also says that at the end of their lives, the most massive stars in the Universe are compelled by gravity to collapse. The star vanishes, leaving an imprint in the fabric of the Universe. But behind a horizon, something remains. A singularity, a moment rather than a place when our knowledge of the laws of Nature breaks down. According to general relativity, the singularity lies at the end of time. There is also a singularity in our past, which marks the beginning of time: the Big Bang. We are asked to accept the profound idea that our scientific description of gravity, the familiar force that governs the behaviour of cannon balls and moons, is at its heart concerned with the nature of space and time.

It's not obvious that gravity should be related to space and time, let alone that seeking to describe it in a scientific theory might lead one to contemplate the beginning and end of time. Black holes take centre stage in exploring this deep relationship because they are gravity's most extreme observable creations. They are so intellectually troublesome that well into the 1960s many physicists felt that while black holes are a feature of the mathematics of general relativity, Nature would surely find a way to avoid creating them. Einstein himself wrote a paper in 1939 in which he concluded that black holes 'do not exist in physical reality'.[2] Einstein's illustrious contemporary Arthur Eddington put it in rather pithier terms: 'There should be a law of Nature to prevent a star from behaving in this absurd way.' Well, there isn't, and they do.

We now understand that black holes are a natural and unavoidable phase in the lives of stars a few times more massive than our Sun, and since there are many millions of such stars in our galaxy, there are many millions of black holes. Stars are large clumps of matter fighting gravitational collapse. In the early stage of their lives, they resist the inward pull of their own gravity by converting hydrogen into helium in their cores. This process, known as nuclear fusion, releases energy which creates a pressure that halts the collapse. Our Sun is currently in this phase, converting 600 million tonnes of hydrogen into helium every second. It's easy to skim over very large numbers in astronomy, but we should pause and marvel at the terrifying difference in scale between the stars and the objects of everyday human experience. Six hundred million tonnes is the mass of a small mountain, and our Sun has been steadily burning through a mountain's-worth of hydrogen every second since before the Earth formed. Not to worry; it has enough hydrogen left to continue its tussle with gravity for another 5 billion years. The Sun can do this because it is big; a million earths would fit comfortably inside. It is 1.4 million kilometres in diameter; a passenger jet would have to fly for six months to circumnavigate it. And yet the Sun is a small star. The largest known stars are a thousand times larger, with diameters in the region of a billion kilometres. Placed at the centre of our solar system, such stars would engulf Jupiter. Monsters like these will end their lives in catastrophic gravitational collapse.

Gravity is a weak but inexorable force. It only attracts, and in the absence of any stronger counteracting forces, it attracts without limit. Gravity is trying to pull you through the floor towards the centre of the Earth, and it's pulling the ground in the same direction. The reason everything doesn't collapse to a central point is that matter is rigid; it's built out of particles that obey the laws of quantum physics and repel each other when they approach too closely. But the rigidity of matter is something of an illusion. We fail to perceive that the ground below us is essentially empty space. The dancing electron clouds surrounding atomic nuclei keep atoms apart and fool us into thinking that solid objects are densely packed. The reality is that the atomic nucleus occupies only a tiny fraction of the volume of an atom and that the ground below our feet is as insubstantial as vapour. The repulsive forces inside matter are nevertheless very powerful and they are capable of keeping you from falling through the floor, and of stabilising dying stars up to twice the mass of the Sun. But there is a limit, and that limit is explored by neutron stars.

A typical neutron star has a radius of just a few kilometres and a mass around 1.5 times that of the Sun. A million 'Earths' squashed into a region the size of a city. Neutron stars tend to spin very fast, emitting bright beams of radio waves that illuminate the Universe like a lighthouse. The first observation of such a neutron star, known as a pulsar, was made by Jocelyn Bell Burnell and Antony Hewish in 1967. So regular is the pulse, which sweeps over Earth every 1.3373 seconds, that Bell Burnell and Hewish christened it Little Green Men-1. The fastest pulsar yet discovered, known as PSR J1748-2446ad, rotates 716 times every second. Neutron stars are extremely energetic celestial objects. On 27 December 2004, a burst of energy hit the Earth, blinding satellites and expanding our ionosphere. The energy was released by the rearrangement of the magnetic field around a neutron star called SGR 1806-20, which lies 50,000 light years from Earth on the other side of the galaxy. In a fifth of a second the star radiated more energy than our Sun emits in a quarter of a million years.

The gravitational pull at the surface of a neutron star is 100 billion times that of Earth. Anything that falls onto the surface is flattened in an instant and transformed into nucleon soup. If you were to fall onto the surface of a neutron star, the particles that were once a part of your voluminous atoms would be transformed into neutrons and squashed together so tightly that they would be jiggling around at near light speed in an attempt to avoid each other. This jiggling can support a neutron star with a mass of around two solar masses, but no more. Beyond this limit, gravity wins. If a little more mass were poured onto its surface, the city-sized star would collapse to form a spacetime singularity. Georges Lemaître, a Catholic priest and one of the founders of modern cosmology, described the Big Bang singularity at the origin of our Universe as a day without a yesterday. A singularity formed by gravitational collapse is a moment with no tomorrow. What remains outside is a dark imprint of what once shone: a black hole.

Today, we have concrete observational evidence that our Universe is populated by black holes. The images shown in Figure 1.1 were obtained by the Event Horizon Telescope Collaboration, a network of radio telescopes located across the Americas, Europe, the Pacific, Greenland and Antarctica. The left-hand image shows the central supermassive black hole in the galaxy M87, which lies 50 million light years from Earth. As is so often the case in science, this fuzzy image from far away grows increasingly wonderful as you learn more about what you are looking at.

This black hole has a mass 6.5 billion times that of our Sun and lies within the dark central region of the image, known as the shadow. This region is dark because gravity is so strong that light cannot escape, and since nothing can travel faster than light,

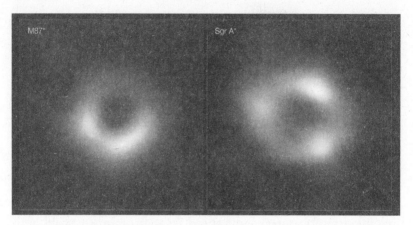

Figure 1.1. Left: The supermassive black hole at the centre of the galaxy M87. Right: Sagittarius A*, the black hole at the centre of our own galaxy. Both as imaged by the Event Horizon Telescope Collaboration. Also on plate 1.

nothing can escape. Inside the shadow lies the event horizon of M87's black hole, a sphere in space of diameter 240 times the distance from the Earth to the Sun. It shields the external Universe from the singularity. The bright disk surrounding the shadow is formed mainly by rays of light emitted from gas and dust spiralling around and into the black hole, their paths twisted and forged into a distinctive doughnut shape by the hole's gravity.

The right-hand image is the supermassive black hole at the centre of our own galaxy, Sagittarius A*. At a mere 4.31 million solar masses, it is a minnow by comparison. The glowing disk would fit comfortably within the orbit of Mercury. Its presence was first inferred indirectly, by observing the orbits of stars around it. These stars are known as the 'S Stars'. The star S2 orbits particularly close to the black hole, with a period of just 16.0518 years. The precision is important, because the detailed observations of S2's orbit were compared with the predictions of general relativity and used to infer the presence of a black hole well before it was photographed. S2 was observed to make its closest approach to Sagittarius A* in 2018, when it passed within just 120

Astronomical Units of the event horizon.* At closest approach, it was travelling at 3 per cent of the speed of light. Reinhard Genzel and Andrea Ghez received the 2020 Nobel Prize for these high-precision observations performed over many years. They were proof that there is a 'supermassive compact object at the centre of our galaxy', in the words of the Nobel Prize committee. They shared the prize with Sir Roger Penrose for his mathematical demonstration 'that black hole formation is a robust prediction of the general theory of relativity'.

We've also detected numerous smaller, stellar mass black holes by detecting the ripples in space and time caused when they collide with each other. In September 2015, the LIGO gravitational wave detector registered the ripples in spacetime caused by a collision between two black holes that occurred 1.3 billion light years from Earth. The black holes were 29 and 36 times the mass of the Sun and collided and merged in less than two tenths of a second. During the collision, the peak power output exceeded that of all the stars in the observable Universe by a factor of 50. By the time the ripples reached us, over a billion years later, they shifted the distance measured along LIGO's 4-kilometre-long laser ruler arms by one thousandth of the diameter of a proton in a fleeting, wiggling pattern that exactly matched the predictions of general relativity. LIGO and its sister detector Virgo have since detected a host of mergers between black holes. The 2017 Nobel Prize in Physics was awarded to Rainer Weiss, Barry Barish and Kip Thorne for their leadership in designing, building and operating LIGO. The known 'Stellar Graveyard' of stellar mass black holes and neutron stars at the time of writing is shown in Figure 1.2.

Taken together, these observations, using different telescopes and techniques, demonstrate beyond reasonable doubt that neutron stars and black holes exist. Science fiction becomes

---

* 1 Astronomical Unit is (approximately) equal to the distance of the Earth from the Sun.

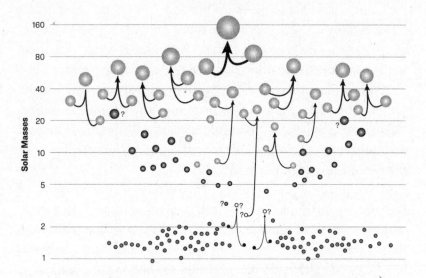

Figure 1.2. The known stellar mass black holes and neutron stars
arranged with the smallest mass objects at the bottom. The smallest
circles are the neutron stars and the arrows indicate observed collisions
and mergers between pairs of black holes or neutron stars.
The numbers on the left are solar masses (mass of
Sun = 1 solar mass).

science when experimental observations confirm theories, and as
our theoretical voyage takes us along ever stranger paths into ever
more tangled intellectual terrain, we should keep reminding
ourselves that these absurd things are real. They are a part of the
natural world, and we should therefore attempt to understand
them using the known laws of Nature. If we fail, we have the
chance to uncover new laws of Nature, and this has most assur-
edly turned out to be the case, beyond even the wildest dreams of
the early pioneers.

Attempting to avoid the absurd

Black holes were first proposed in 1783 by the English rector and scientist John Michell and independently in 1798 by the French mathematician Pierre-Simon Laplace. Michell and Laplace reasoned that, just as a ball thrown upwards is slowed down and pulled back to the ground by the Earth's gravity, it is conceivable that there exist objects that exert such a strong gravitational pull they could trap light.

An object flung upwards from the surface of the Earth must have a speed in excess of 11 kilometres per second to escape into deep space. This is known as Earth's escape velocity. The gravitational pull at the Sun's surface is much stronger, and the escape velocity is correspondingly higher at 620 kilometres per second. At the surface of a neutron star, the escape velocity can approach an appreciable fraction of the speed of light.* Laplace calculated that a body with a density comparable to the Earth but with a diameter 250 times larger than the Sun would have a gravitational pull so great that the escape velocity would exceed the speed of light, and therefore 'the largest bodies in the Universe may thus be invisible by reason of their magnitude'.[3] This was a fascinating idea and ahead of its time. Imagine a spherical shell in space touching the surface of one of Laplace's giant dark stars. The escape velocity from the shell would be the speed of light. Now make the star a little denser. The stellar surface would shrink inwards, but the imaginary shell would remain in place, marking out a boundary in space. If you hovered on the shell, now above the surface of the star, and shone a torch outwards, the light would go nowhere. It would remain forever frozen, unable to escape. This boundary is the event horizon. Inside the shell, the torchlight would be turned around and pulled back onto the star. Only outside of the shell could light escape.

---

* The speed of light is 299,792,458 metres per second.

Michell and Laplace imagined these dark stars as huge objects, perhaps because they could not conceive of the alternative. But an object doesn't have to be big to have a strong gravitational pull at its surface. It can also be very small and very dense; a neutron star, for example. For an object of any mass, one can use Isaac Newton's laws to calculate the radius of the region of no escape that would form around it if it were compressed sufficiently:

$$R_S = \frac{2GM}{c^2}$$

where G is Newton's gravitational constant, which encodes the strength of gravity, and c is the speed of light. If we crush anything with mass M into a ball smaller than this radius, we will have created a dark star. Putting the mass of the Sun into this equation, we find that the radius is approximately 3 kilometres. For the Earth, it's just under 1 centimetre. It is difficult to imagine the Earth being crushed to the size of a pebble, which is probably why Michell and Laplace didn't consider the possibility. Fantastical as they are, however, there would seem to be nothing particularly troublesome or absurd about dark stars, should they exist. They would trap light but, as Laplace pointed out, that would just mean that we wouldn't be able to see them.

This simple Newtonian argument gives us a feel for the idea of a black hole – gravity can get so strong that light cannot escape – but Newton's law of gravitation is not applicable when gravity is strong and Einstein's theory must be used. General relativity also allows for objects whose gravitational pull is so great that light cannot escape, but the consequences are very different and most definitely troublesome and absurd. As in the Newtonian case, if any object is compressed below a certain critical radius, it will trap light. In general relativity this radius is known as the Schwarzschild radius, because it was first calculated in 1915, very shortly after general relativity was published, by the German physicist

Karl Schwarzschild. Coincidentally, the expression for the Schwarzschild radius in general relativity is precisely the same as the Newtonian result above. The Schwarzschild radius is the radius of the event horizon of a black hole.

We will learn more about the Schwarzschild radius in Chapter 4, when we have the machinery of general relativity at our disposal, but we can catch a glimpse of some of the absurdities to come. We will learn that black holes affect the flow of time in their vicinity. As an astronaut falls towards a black hole, their time will tick more slowly as measured on clocks far away in space. That's interesting, but not absurd. The absurd-sounding result is this: according to the far away clocks, time grinds to a halt on the event horizon. As viewed from the outside, nothing is ever seen to fall into a black hole, which means an astronaut falling towards a black hole will remain frozen on the horizon for all eternity. This also applies to the surface of a star collapsing inwards through the horizon to form the black hole. At first sight, it seems the theory of general relativity predicts a nonsense. How can a star collapse through the event horizon to form a black hole if its surface is never seen to cross the horizon? Observations like this troubled Einstein and the early pioneers, and this is only one in a blizzard of apparent paradoxes.

For Einstein, and the majority of physicists until the 1960s, such worries led to the conclusion that Nature would find a way out, and research into black holes was primarily concerned with demonstrating that they could not exist. Perhaps it is not possible to compress a star without limit and thereby generate an event horizon. This doesn't seem unreasonable, given that a sugar-cube-sized lump of neutron star material would weigh at least 100 million tonnes. Perhaps we don't fully understand how matter behaves at such extreme densities and pressures.

Stars are large clumps of matter fighting gravitational collapse, and when they run out of nuclear fuel their fate depends on their mass. In 1926, Eddington's Cambridge colleague R. H. Fowler published an article 'On Dense Matter' in which he showed that the newly-discovered quantum theory provided a way for an old collapsing star to avoid forming an event horizon due to an effect known as 'electron degeneracy pressure'.[4] This was the first glimpse of the 'quantum jiggling' we referred to earlier in the context of neutron stars. His conclusion appeared to be an unavoidable consequence of two of the cornerstones of quantum theory: Wolfgang Pauli's Exclusion Principle and Werner Heisenberg's Uncertainty Principle.

The Exclusion Principle states that particles like electrons cannot occupy the same region of space. If lots of electrons are squashed together by gravitational collapse, they will separate themselves into their own individual tiny volumes inside the star in order to stay away from each other. Heisenberg's Uncertainty Principle now comes into play. It states that as a particle is confined to a smaller volume, its momentum becomes larger. In other words, if you confine an electron it will jiggle around, and the more you try to confine it, the more it will jiggle. This creates a pressure in much the same way that the heat from nuclear fusion reactions earlier in the star's life causes its atoms to jiggle and halt the collapse. Unlike the pressure from fusion reactions, however, electron degeneracy pressure requires no energy release to power it. It seemed a star could resist the inward pull of gravity indefinitely.

Astronomers knew of such a star, known as a white dwarf. Sirius B is a faint companion of Sirius, the brightest star in the heavens. Sirius B was known to have a mass close to that of our Sun, but a radius comparable to the Earth. Its density, using the measurements of the time, was estimated to be around 100 kg/cm^3, which, as Fowler notes 'has already given rise to most interesting theoretical considerations'. In his book *The Internal Constitution of Stars*, Eddington wrote, 'I think it is generally considered proper to add the conclusion "which is absurd".' Modern measurements put the density over ten times higher. Absurd as this exotic planet-sized star appeared, however, Fowler had discovered a mechanism that explained how it could resist gravity. This seems to have offered a great deal of relief to the physicists of the day because it stopped the unthinkable happening. Thanks to Fowler, it appeared that stars end their lives as white dwarfs. Supported by the quantum jiggling of electrons, they will not collapse inside the Schwarzschild radius and an event horizon will not form.

The sense of relief didn't last long. In 1930, during an 18-day voyage from Madras to work with Eddington and Fowler in Cambridge, a 19-year-old physicist named Subrahmanyan Chandrasekhar decided to calculate just how powerful electron degeneracy pressure could be. Fowler had not placed an upper limit on the mass of a star supported in this way, and it seems most physicists assumed there should be none. But Chandrasekhar realised that electron degeneracy pressure has its limits. Einstein's theory of relativity says that no matter how confined an electron becomes, the speed of its jiggles cannot exceed the speed of light. Chandrasekhar calculated that the speed limit will be reached for a white dwarf with a mass around 90 per cent that of the Sun.[5] A more accurate calculation reveals that the Chandrasekhar limit, as it is now known, is 1.4 times the mass of the Sun. If a collapsing star exceeds this mass, the electrons no longer provide enough pressure to resist the inward pull of gravity because they are moving as fast as they can, and gravitational collapse must continue. Eddington was unimpressed. He felt that Chandrasekhar had incorrectly meshed relativity with the then-new field of quantum mechanics and, when done correctly, the calculation would show that white dwarf stars could exist up to arbitrarily large masses. The subsequent argument between the young Chandrasekhar and the venerable Eddington affected Chandrasekhar deeply. Decades after Eddington's death in 1944, Chandrasekhar still described this time as 'a very discouraging experience … to have my work completely discredited by the astronomical community'. Chandrasekhar was ultimately proved correct, and he received the Nobel Prize for his work on the structure of stars in 1983.

Chandrasekhar's result, published in 1931, was not regarded as definitive evidence that black holes must form. Einstein was still concerned about the apparent freezing of time on the event horizon in 1939. Perhaps there is some other process that can provide support for a collapsing white dwarf when electron degeneracy pressure fails? In the late 1930s, the American physicist Fritz Zwicky and the Russian physicist Lev Landau suggested, correctly, that there may be even denser stars than white dwarfs that are supported not by electron degeneracy pressure but by neutron degeneracy pressure. Under the extreme conditions found in gravitational collapse, electrons can be forced to fuse together with protons to form neutrons and lightweight particles called neutrinos, which escape the star. Neutrons, just like electrons, jiggle around as they are squashed close together, but because they are more massive than electrons, they can provide more support. These objects are neutron stars.

It's not unreasonable to wonder whether this fate might be the end of the road for all supermassive stars, even though the experience with white dwarfs suggests that neutron degeneracy pressure should also have its limits. Maybe the most massive stars eject material into space as they collapse, or maybe they bounce and explode as they reach neutron star densities. These possibilities were not easily dismissed at the time – nuclear physics was a very new field and the neutron itself was only discovered in 1932.

By 1939, J. Robert Oppenheimer and his student George Volkov, building on work by Richard Tolman, had established what is now called the Tolman–Oppenheimer–Volkov limit, which places an upper limit on the mass of a neutron star at just under three times the mass of the Sun. Oppenheimer and another of his students, Hartland Snyder, subsequently showed that, under certain assumptions, the heaviest stars *must* collapse behind an event horizon to form a black hole.[6] This landmark paper begins: 'When all thermonuclear sources of energy are exhausted a sufficiently heavy star will collapse. Unless fission due to rotation, the radiation of mass, or the blowing off of mass by radiation, reduce the star's mass to the order of that of the Sun, this contraction will continue indefinitely.' The final lines of the introduction detail the consequences for the flow of time at the horizon that so worried Einstein: 'The total time of collapse for an observer comoving with the stellar matter is finite, and for this idealized case and typical stellar masses, of the order of a day; an external observer sees the star asymptotically shrinking to its gravitational radius.'* In other words, it takes around a day for a star not very much bigger than the Sun to collapse out of existence from the point of view of someone riding inwards on the surface of the collapsing star, but an eternity for anyone watching from the outside. This is the puzzling behaviour of time we noted previously. Oppenheimer and Snyder accepted this basic result of general relativity and showed that it leads to no contradiction. We will explore these intriguing results in more detail in the following chapters.

At this point, World War II intervened, and the thoughts of the world's physicists turned to supporting the war effort. In the United States, the expertise in nuclear physics honed by the study of stars was particularly relevant to the development of the atomic bomb, and Oppenheimer famously became the scientific leader of the Manhattan Project. When the war ended and the physicists returned, a new generation was poised to take up the mantle. In the United States, that generation was nurtured by John Archibald Wheeler. It was Wheeler who first coined the term black hole at a lecture in the West Ballroom of the New York Hilton on 29 December 1967. In his autobiography, Wheeler describes his intellectual struggles with black holes throughout the 1950s.[7] 'For some years this idea of collapse to what we now call a black hole

---

* By 'gravitational radius' they mean the Schwarzschild radius.

went against my grain. I just didn't like it. I tried my hardest to find a way out, to avoid compulsory implosion of great masses.' He recounts how he eventually became convinced that 'nothing can prevent a large-enough chunk of cold matter from collapsing to a dimension smaller than the Schwarzschild radius'. Wheeler's intellectual conversion culminated in a 1962 paper with his student Robert Fuller in which they conclude that 'there exist points in spacetime from which light signals can never be received, no matter how long one waits'.[8] These are the points inside the event horizon from which the Universe beyond is forever isolated. Black holes, it seems, are unavoidable. Any remaining theoretical concerns were dispelled in 1965 by Sir Roger Penrose's Nobel Prize-winning paper 'Gravitational Collapse and Space-Time Singularities', a three-page tour-de-force in which Penrose proves that, in Wheeler's words, 'for just about any description of matter that anyone has imagined, a singularity must sit at the centre of a black hole'.[9]

A profound glow

Our brief history of black holes brings us to 1974 and a paper by Stephen Hawking, which led to an apparently simple question that has driven black hole research for half a century since its publication.

By the 1970s the existence of black holes was widely accepted by theorists, although they were yet to be sighted by astronomers, and the attention of the small group still interested in them turned to the conceptual challenges they pose. Hawking's paper, published in the journal *Nature*, is colourfully titled 'Black Hole Explosions?'[10] Hawking showed that the presence of an event horizon has a dramatic effect on the vacuum of space in its vicinity. Quantum theory tells us that empty space is not empty. It is filled with fields that are constantly fluctuating, and these fluctuations manifest themselves as the potential to create particles:

photons, electrons, quarks, any particles, in fact. The vacuum has a structure. In common or garden empty space, these fluctuations come and go; one might picture so-called virtual particles continually popping into and out of existence, but the net result is that no real particles ever appear miraculously out of nothing. The presence of the horizon disrupts this balance, with the result that the fleeting virtual particles can become real. These particles, known as Hawking radiation, stream out into the Universe carrying a tiny fraction of the black hole's energy with them. Over unimaginable time scales, vastly longer than the current age of the Universe, a typical black hole will evaporate away and, ultimately, explode. Black holes, to use Hawking's famous phrase, ain't so black. They glow gently like faint coals in the cold sky. Very faint coals. The temperature of a solar mass black hole is 0.00000006 degrees Celsius above absolute zero, which is far colder than the Universe today.* Sagittarius A* is even colder: 4.31 million times colder to be precise. But the temperature of a black hole is not zero, and that matters enormously. It means, as we'll discover, that black holes obey the laws of thermodynamics – the same laws that govern glowing coals and steam engines and stars – and it means they are not immortal. One day in the far, far future, they will all be gone.

A profound question arises as a result of this faint glow. When the black hole has gone, what has happened to everything that fell in? Because of the unique production mechanism of Hawking radiation, plucked as it is out of the vacuum in the vicinity of the event horizon, the radiation would seem to have nothing to do with whatever has fallen into the black hole during its lifetime. It is very difficult therefore to see how any information about anything that fell in, or indeed the star that collapsed to form the black hole in the first place, could be preserved, imprinted some-

---

* The temperature of the cosmic microwave background radiation today is 2.725 degrees Celsius above absolute zero.

how, in the radiation. Indeed, Hawking's original calculation appeared very clear on this point. The radiation, the remnants of the black hole, contains no information at all.

One of the pioneers of modern black hole research, Leonard Susskind, tells the story of a meeting in a small San Francisco attic room in 1983 at which Hawking first raised this question and answered it, incorrectly as it turns out. Susskind's first-hand account of the tremendous intellectual struggle Hawking's question generated is called *The Black Hole War: My Battle with Stephen Hawking to Make the World Safe for Quantum Mechanics*. Susskind has a way with titles. He once co-authored a paper called 'Invasion of the Giant Gravitons from Anti-de Sitter Space'. He writes that 'Stephen claimed that information is lost in black hole evaporation, and worse, he seemed to prove it. If that was true ... the foundations of our subject were destroyed.'

Susskind was referring to one of the pillars of modern physics: determinism. If we know everything about a system, be it a simple box of gas or the Universe, we can predict how it will evolve into the future and how it looked in the past. This is an 'in principle' statement of course. It's not possible in practice to know everything about the past and future, because we always have incomplete information about any real physical system. But in science, unlike modern-day politics, principles matter. If Hawking was right, black holes would render the Universe fundamentally unpredictable and the foundations of physics would crumble.

We now know that Stephen Hawking was wrong – information is not destroyed and physics is safe – as Hawking himself came to accept with delight, not regret, not least because the ongoing programme of research stimulated by his original claim continues to propel us towards a new understanding of space and time and the nature of physical reality.

In the last edition of *A Brief History of Time*, Hawking writes that he eventually changed his mind in 2004 and conceded a bet he'd made with John Preskill (whose work we'll meet later). After a further argument about the merits of cricket and baseball, which he also lost, Hawking gave Preskill an encyclopaedia of baseball. At the time of writing, Hawking notes, nobody knew how the information gets out of the black hole – just that it does. What was clear, however, is that the information would be very hard to decode. 'It's like burning a book,' he writes. 'Information is not technically lost, if one keeps the ashes and the smoke – which makes me think again about the baseball encyclopaedia I gave John Preskill. I should perhaps have given him its burnt remains instead.'

Beyond the horizon

Imagine you find a watch lying on the ground. On close inspection you are compelled to marvel at its delicate sophistication and exquisite precision. The mechanism was surely designed; there must have been a creator. Transpose 'watch' for 'Nature' and this is the argument for God presented by clergyman William Paley in 1802. We now understand that the argument is seriously undermined by the overwhelming evidence in support of Darwin's theory of evolution by natural selection. The watchmaker is Nature, and it is blind. 'There is grandeur in this view of life,' wrote Darwin, 'with its several powers, having been originally breathed into a few forms or into one; and that, while this planet has gone cycling on according to the fixed law of gravity, from so simple a beginning endless forms most beautiful and most wonderful have been, and are being, evolved.'

But what of the fixed law of gravity, a prerequisite for the existence of the planets on which the endless forms evolved? Or the laws of electricity and magnetism which glue the animals together? Or the menagerie of subatomic particles out of which we are made? Who or what laid down the laws; the framework within which everything cycles on?

The story of modern physics has been one of reductionism. We do not need a vast encyclopaedia to understand the inner workings of Nature. Rather, we can describe a near-limitless range of natural phenomena, from the interior of a proton to the creation of galaxies, with apparently unreasonable efficiency using the language of mathematics. In the words of theoretical physicist Eugene Wigner, 'The miracle of the appropriateness of the language of mathematics for the formulation of the laws of physics is a wonderful gift which we neither understand nor deserve. We should be grateful for it.'[11] The mathematics of the twentieth century described a Universe populated by a limited number of different types of fundamental particles interacting with each other in an arena known as spacetime according to a collection of rules that can be written down on the back of an envelope. If the Universe was designed, it seemed, the designer was a mathematician.

Today, the study of black holes appears to be edging us in a new direction, towards a language more often used by quantum computer scientists. The language of information. Space and time may be emergent entities that do not exist in the deepest description of Nature. Instead, they are synthesised out of entangled quantum bits of information in a way that resembles a cleverly constructed computer code. If the Universe is designed, it seems, the designer is a programmer.

But we must take care. Like Paley before us, we are in danger of over-reaching. The role of information science in describing black holes may be pointing us towards a novel description of Nature, but this does not imply we were programmed. Rather we might conclude that the language of computing is well suited to describing the algorithmic unfolding of the cosmos. Put in these terms, there is no greater or lesser mystery here than Wigner's miracle of the appropriateness of the language of mathematics for the formulation of the laws of physics. Information processing – the churning of bits from input to output – is not a construction of computer science, it is a feature of our Universe. Rather than spacetime-as-a-quantum-computer-code pointing to a programmer, we might instead take the view that earth-bound computer scientists have discovered tricks that Nature has already exploited. Viewed in this way, black holes are cosmic Rosetta Stones, allowing us to translate our observations into a new language that affords us a glimpse of the profoundest reason and most radiant beauty.

2

UNIFYING SPACE AND TIME

'The word "distance" by itself does not belong in a book on general relativity. The word "time" by itself does not belong in a book on general relativity.'

Edwin F. Taylor, John Archibald Wheeler
and Edmund Bertschinger[12]

Black holes are perfect for learning about physics because understanding them requires pretty much all of it. Don Page begins his exhaustive 'inexhaustive review of Hawking radiation' with the sentence: 'Black holes are perhaps the most perfectly thermal objects in the universe, and yet their thermal properties are not fully understood.'[13] Thermodynamics is one of the cornerstones of physics, dealing with familiar concepts such as temperature and energy, and a possibly less-familiar concept, entropy. Thus, we will need to learn some thermodynamics. Stephen Hawking's seminal paper 'Particle Creation by Black Holes' begins: 'In the classical theory black holes can only absorb and not emit particles. However it is shown that quantum mechanical effects cause black holes to create and emit particles as if they were hot bodies ...'[14] Thus, we will need to learn some quantum mechanics. And, of course, there is Einstein's General Theory of Relativity, wherein, as Misner, Thorne and Wheeler write in their great (in quality and in mass) textbook *Gravitation*, '... the reader is transported to the land of black holes, and encounters colonies of static limits, ergospheres, and horizons – behind whose veils are hidden gaping, ferocious singularities'.[15] This is the land we will explore first.

We learn at school that gravity is a rather mundane thing – the force between everyday objects; you can't jump too high from the surface of the Earth because there is a force that pulls you back down to the ground. In 1687, Isaac Newton formalised this idea and published it in *The Principia Mathematica*. Newton's theory works well in most situations, allowing us to calculate the trajectories of spacecraft to the Moon and beyond, and at first sight has nothing to say about space and time at all. Newton did, however, assume two properties of space and time in formulating the theory. He assumed that time is universal: if everyone in the Universe carries a perfect clock and all the clocks were synchronised sometime in the past, they will all read the same time in the future. Newton put it more poetically: 'Absolute, true and mathematical time, of itself, and from its own nature flows equably without regard to anything external ...' He also assumed that space is absolute: a great arena within which we live out our lives. 'Absolute space, in its own nature, without regard to anything external, remains always similar and immovable ... Absolute motion is the translation of a body from one absolute place into another.' These assumptions sound like common sense – so much so that it's a testament to Newton's genius that he even noticed he'd made them. His true genius is revealed when we discover that his care was prescient because both assumptions are wrong. The Universe is not constructed this way, and as the foundations of the theory crumble, so must the theory itself. Einstein's General Theory of Relativity is the replacement, describing a Universe in which distances in space and the rate at which time ticks depend on an observer's proximity to stars and planets and black holes or even on their route to the shops and back.

It is an experimental fact that the passage of time varies from place to place and depends on how fast things move relative to each other. In a wonderfully simple experiment, carried out in 1971, Joseph C. Hafele and Richard E. Keating bought round-the-world airline tickets for themselves and four high-precision atomic clocks. In their own carefully chosen words: 'In science, relevant experimental facts supersede theoretical arguments. In an attempt to throw some empirical light on the question of whether macroscopic clocks record time in accordance with the conventional interpretation of Einstein's relativity theory, we flew four caesium beam atomic clocks around the world on commercial jet flights, first eastward, then westward. Then we compared the time they recorded during each trip with the corresponding time recorded by the reference atomic time scale at the US Naval Observatory. As was expected from theoretical predictions, the flying clocks lost time (aged slower) during the eastward trip and gained time (aged faster) during the westward trip.'[16] The eastward clocks lost 59 nanoseconds and the westward clocks gained 273 nanoseconds.* These are tiny time differences over such a long journey, but they are not zero and, most importantly, the experimental observations agree with the mathematical calculations performed using Einstein's theory. The Hafele–Keating paper finishes in a similarly concise fashion: 'In any event, there seems to be little basis for further arguments about whether clocks will indicate the same time after a round trip, for we find that they do not.' And there we have it – a remarkable and highly unexpected feature of our Universe that relativity theory is designed to describe: time is not what it seems.

Space is not what it seems either: in a further affront to common sense, the distance between two points in space is not something everyone will agree upon. Hold your fingers apart in front of you. Who would dare say that the distance between your fingertips depends on the point of view? Einstein would. This is

---

* A nanosecond is one billionth of a second.

also a well-verified experimental fact. The Large Hadron Collider at CERN is the world's most powerful particle accelerator. The giant machine's job is to make protons travel around its underground tunnel at 99.999999 per cent the speed of light, before smashing them together. The purpose is to explore the structure of matter and the forces of Nature that animate our world. The LHC is 27 kilometres in circumference from the point of view of someone standing on the ground in Geneva, marvelling at this great engineering achievement. From the point of view of the protons orbiting around the ring, the circumference is 4 metres.

Einstein didn't know about atomic clocks or airliners or the Large Hadron Collider in 1905, and no experiments had been performed that challenged Newton's pictures of absolute space and universal time. Why, then, did Einstein decide to invent a new picture? The answer is that he realised there is a fundamental clash between Newton's seventeenth-century theory of gravitation and James Clerk Maxwell's nineteenth-century theory of electricity and magnetism.

The clash concerns the way the speed of light appears in Maxwell's theory. The theory, which is based on experimental observations carried out by Michael Faraday, André-Marie Ampère and others throughout the nineteenth century, states that light is an electromagnetic wave that travels through the vacuum of empty space at a fixed speed: 299,792,458 metres per second. According to the theory, the speed of a beam of light is always this precise number, no matter how the person that measures it moves relative to the source of the light. That's a very strange prediction, and not the way most other things in Nature behave.

The fastest ball ever bowled in international cricket, at the time of writing, was by Shoaib Akhtar for Pakistan against England in Cape Town in 2003. Nick Knight, opening for England, played a textbook defensive stroke to square leg, rounding off a maiden over for Akhtar. The ball travelled down the wicket at 100.2

mph.* If Akhtar had instead bowled the ball from a Grumman F14 Tomcat travelling at 600 mph directly towards Knight, then the ball would have reached the batsman at 600 + 100.2 = 700.2 mph, and he may not have guided it to square leg. This is not true for light. If, rather than the cricket ball, a laser beam had been sent towards Knight from the F14 Tomcat, the light would still have reached him at the speed of light (not the speed of light + 600 mph).

There are two possible resolutions to this bizarre feature of Maxwell's equations. The obvious one would be to modify Maxwell's equations so that this is not the case, and light behaves like a cricket ball. Ultimately this is an experimental question; a question about what actually happens in Nature. Innumerable observations of disparate physical phenomena for well over a hundred years tell us that Maxwell's equations are correct as they stand and therefore light always travels at the same speed.

The other, less obvious, resolution is to change the way that observers travelling at different speeds relative to each other account for distances and time differences such that everybody always measures the speed of light to be the same. Einstein chose this route, thus rejecting Newton's notions of absolute space and time, and this choice led him to relativity.

Einstein's theory of relativity

Einstein's theory is a model, which is to say it is a mathematical framework that allows us to make predictions about how objects that exist in the natural world behave. The model is inherently geometrical, which lends itself to intuitive visual pictures which require very few equations – a good thing for a book such as this. We believe that the best approach to explaining relativity is to describe this geometrical picture, rather than attempt to present

---

* We will use imperial units when discussing cricket.

its evolution historically. Our justification is that the model works, and that is the only justification necessary. Einstein could have simply plucked his theory out of thin air without any reference to Maxwell's theory or experiments, and it would be equally valid because it is a good model in the sense that its predictions have passed every experimental test to date.

If Einstein could have plucked one single idea out of thin air that would have led him directly to his theory, including the explanation of what happened in Hafele and Keating's experiment and the most famous equation in all of physics, $E = mc^2$, it would be a concept known as 'the spacetime interval'. The idea is beautifully simple.

Let's return to Pakistan against England in Cape Town and Shoaib Akhtar's record-breaking delivery to Nick Knight. We are going to simplify things for now by switching gravity off – we'll switch it back on at the end of this chapter. This means that when the ball leaves Akhtar's hand it will travel to Knight in a perfect straight line at a constant speed – 100.2 mph relative to the ground.* Let's further imagine that the cricket ball has a clock inside. At the moment the ball leaves Akhtar's hand, the ball emits a flash of light and records the time on its internal clock. At the instant the ball reaches Knight's bat, the ball emits another flash of light and records the time of arrival on its internal clock. We'll call the time interval between the flashes as measured on the cricket ball clock $\Delta\tau$ – pronounced delta tau.

In the commentary box, Jonathan Agnew (Aggers), for the BBC, notes the arrival of the two flashes of light from the ball and calculates the time interval between the emission of the flashes from his point of view: Δt_{Aggers}.† He also measures the distance

---

* In more technical language, we are assuming that the cricket ground is an inertial reference frame. We might imagine it detached from Earth and floating freely between the stars. We are also neglecting air resistance.
† He'll need to make a correction for the time it takes for the light to travel from the ball to his eyes in order to work out when the flash was actually emitted.

between the place where the ball leaves Akhtar's hand and the place where the ball hits Knight's bat: Δx_{Aggers}.

In his Grumman F14 Tomcat flying over the wicket in a straight line between the stumps at 600 mph Tom, the pilot, also notes the two flashes of light and calculates the time interval between the emission of the flashes from his point of view: Δt_{Tom}. Like Aggers, he also measures the distance between the place where the ball leaves Akhtar's hand and the place where the ball hits Knight's bat: Δx_{Tom}.

The Hafele and Keating result tells us that the time differences between the emission of the flashes as measured by Aggers, Tom and the cricket ball will all be different. Likewise, the distance the ball travelled from bowler to batsman will also be different. For those who have never encountered Einstein's ideas before, these differences should come as a tremendous shock. They are counter-intuitive because they mean that distances and time intervals are not something everyone can agree upon. However, here is a remarkable and important result. If Aggers calculates the quantity $(\Delta t_{\text{Aggers}})^2 - (\Delta x_{\text{Aggers}})^2$ and Tom calculates the quantity $(\Delta t_{\text{Tom}})^2 - (\Delta x_{\text{Tom}})^2$ then they will both get the same result, and the result will be equal to the square of the time interval measured using the cricket ball clock, $(\Delta \tau)^2$:

$$(\Delta \tau)^2 \;=\; (\Delta t_{\text{Aggers}})^2 \;-\; (\Delta x_{\text{Aggers}})^2 \;=\; (\Delta t_{\text{Tom}})^2 \;-\; (\Delta x_{\text{Tom}})^2$$

$(\Delta \tau)^2$ is known as the spacetime interval between the two events: event 1 is the ball leaving the bowler's hand and event 2 is the ball striking the bat. You may well ask: 'What does it mean to subtract a distance in space squared from a time difference squared?' The answer is that we must specify the distance between two events as the time it takes for light to travel between those events, which means we should compute the distance in light seconds. The spacetime interval (or 'interval' for short) is important because it is a quantity on which everyone agrees, no matter what their point of view. In physics we call such a quantity an invariant. Since Nature doesn't care about our point of view,* we should only seek to describe Nature in terms of invariant quantities. When we discover an invariant, it is a big deal because we learn a little more about the essential structure of the Universe.

In their book *Exploring Black Holes*, Taylor, Wheeler and Bertschinger describe the equation for the interval as 'one of the greatest equations in physics, perhaps in all of science'. Kip Thorne and Roger Blandford, in *Modern Classical Physics* write that the interval is 'among the most fundamental aspects of physical law'. The word 'fundamental' is important. You might reasonably ask: 'Why is the interval like this?' 'Why does everyone agree on this particular combination of time and space?' The answer, as Thorne and Blandford imply by their use of the word 'fundamental', is that this is the way the Universe is constructed. We know of no deeper explanation for the form of the interval.

A further question you may be asking is: 'How should I think about the interval, this most fundamental aspect of physical law?' This is a good question. Physicists usually endeavour to develop a mental picture of what's happening in their equations: physical intuition brings equations to life. Fortunately, the interval does have a simple physical interpretation. It is related to what we will refer to as 'the distance between two events'. This is not the usual distance between them in space, but the distance in spacetime. Let's explore that idea.

Events and worldlines in spacetime

The concept of an event is fundamental to relativity. An event is something that happens somewhere and somewhen. You clicking your fingers is a good approximation of an event: it happens very

---

* This comes as a shock to a certain type of individual.

quickly and in a well-defined location. The emission of a flash of light from our cricket ball is an event. Strictly speaking an event is an idealised concept, something that happens so fast and in such a small area that it corresponds to a single point in space and time. The theory of relativity is concerned with the relationship between events; how far apart they are in spacetime and whether they influence each other or not. This is a very intuitive way to think about the world, so much so that it's how we speak in everyday life; 'I'll meet you tomorrow evening at eight o'clock at the pub.' 'I was born on 3 March 1968 in Oldham.' Things that have happened to us and things that will happen to us are all events in space and time, and they happen somewhere and somewhen. A slight shift in wording, and we have the basis of the theory of relativity: things that have happened to us and things that will happen to us are all events in spacetime. What is spacetime? It's the collection of all events. Everything that has ever been and will ever be in the Universe.

Here is a picture of spacetime. Picture the events of your life. Your first day at school. Christmas with your grandparents. That night down the pub. A chronicle of moments from joy to despair and everything in between. Events are the atoms of experience. From our human perspective, events come with labels; we speak of them in terms of the place and time they happen. Imagine carefully laying out the events of your life one by one to form a line snaking over spacetime; an unbroken path charting your journey through the world. This is called your worldline.

Figure 2.1 depicts a worldline winding its way over spacetime. It is known as a spacetime diagram. Imagine this is your worldline, your life laid out before you. Change the dates, add your own events and memories, construct the map of your experiences. Spacetime is an evocative thing. The collection of all events in your life, past, present and future. Your memories are of events in spacetime. The moments that make up your life – Christmases long ago, summer afternoons with school-friends, first kisses and

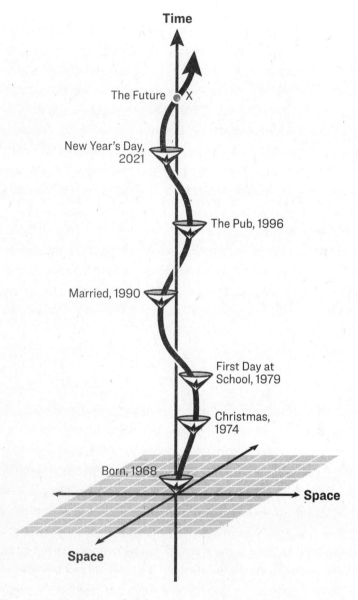

Figure 2.1. The events of a life in space and time. The line through the events is known as a worldline. The cones at each event are known as light cones. They are the paths of a flash of light, emitted at the event. Because nothing can travel faster than light, only future events inside these cones can be influenced by the original event.

last goodbyes – have not been lost forever. Those moments are still out there, somewhere in spacetime. Your future – everything that has yet to happen to you – every event including your death at the end of your worldline – is waiting for you to arrive, somewhere in spacetime. If we lay out all events in this way, we have created a map of spacetime, and the distance between the events is given by the interval. How wonderful it would be to have freedom of movement over this map, the ability to revisit every moment. We can move anywhere over a map of space, so why should we not have the same freedom over a map of spacetime? The reason is to be found in the interval.

Let's recap. From a particular point of view, the distance in space between two events is measured to be Δx and the difference in time between the events is measured to be Δt. From a different point of view, Δx and Δt will be different, which is very coun-ter-intuitive. But crucially, the interval $(\Delta \tau)^2$ will not depend on point of view:

$$(\Delta \tau)^2 \; = \; (\Delta t)^2 \; - \; (\Delta x)^2$$

We can use the idea of the interval to introduce the concept of the length of a worldline. To be specific, think of the worldline in Figure 2.1 as it goes from being born in 1968 to the enigmatic future event marked X. How long is this portion of the worldline? If event X occurs at precisely the same location as the birth, then the equation above informs us that the interval between the two events (birth and X) is just given by the time interval, i.e. $\Delta \tau = \Delta t$ because $\Delta x = 0$. This is the interval between the two events, but it is not the length of the worldline. Rather it is the length of the worldline that goes straight up the time axis (the vertical line on Figure 2.1). Just like the distance of a journey between Oldham and Wigan depends on the route taken, so it is with distances in spacetime. They depend on the spacetime path taken by the worldline. The way to compute the length of the snaking worldline in Figure 2.1 is to imagine chopping it into lots of tiny segments. Each segment being approximately a straight line.* Then we can compute $\Delta\tau$ for each segment, using the formula above, and add up all the $\Delta\tau$'s to get the total length.

We can also make the important observation that there are three different sorts of interval: $(\Delta\tau)^2$ can be positive, negative or zero. We might say that there are three different sorts of 'distance' in spacetime, in contrast to one sort of distance in space.

If the time difference between the events is larger than the distance in space between them, the interval is positive. Such pairs of events are referred to as 'timelike separated'. All the events on your worldline are timelike separated from each other. There is a simple physical interpretation for the interval in this case. If you had a perfect stopwatch that you started at the moment you were born, and carried it with you for your whole life, the watch would measure the length of your worldline, from your birth to the present moment. The length of your worldline is therefore your age. This is the meaning of the interval for timelike separated events. It is the time measured on a watch carried along a worldline between events.†

If the distance in space between the events is larger than the time difference, the interval is negative. We say these events are 'spacelike' separated. We can now no longer interpret the interval in terms of a watch moving between the events. A physical interpretation does exist, however. For the case where the two events occur at the same time, we can interpret the interval as recording the distance between these events as measured on a ruler. It turns out that for spacelike separated events, it is always possible to find an observer (i.e. a point of view) from whose perspective the

---

* This means that the person travelling along the worldline is not accelerating during the segment. Any curving path, through space or through spacetime, can be thought of as being made up of lots of tiny straight-line paths.

† To see this, notice that the watch never moves relative to itself, so $\Delta\tau = \Delta t$ because $\Delta x = 0$.

events happen at the same time. This means it's not possible for someone or something to be physically present at both events since that would require being in two places at the same time. That's just another way of saying that we could not arrange to carry a watch between the events.

There are therefore two fundamentally different regions of spacetime surrounding any event: a region containing those events that could conceivably be on the worldline of a watch passing through that event, and a region containing events that could not. We'll see the significance of this division in a moment.

The third possibility is that the time difference between a pair of events is exactly equal to the distance in space between them. This is the case if a worldline between the two events is the path taken by a beam of light. To see this, recall that we measure time in seconds and distance in light seconds. Light travels 1 light second in 1 second, 2 light seconds in 2 seconds, and so on. So, for any pair of events that lie on the path of a beam of light, $(\Delta t)^2 = (\Delta x)^2$ and the interval is zero. These events are known as 'light-like' separated. If we draw the paths of light rays out over spacetime from an event, they form what is known as the future light cone of the event. In Figure 2.1, the light cones are depicted as small cones at each event. The light cones spread out at an angle of 45 degrees from each event. Inside the future light cone, all events are timelike separated from the original event while outside the future light cone, all events are spacelike separated from the original event. Since we were present at every event in our own lives, our worldline snakes along inside the light cones.*

It is important to understand the meaning of the light cones and what they tell us about the relationships between events in

---

* Light cones are cones in spacetime if space is two-dimensional because a flash of light will spread out in a circle of increasing radius. In three dimensions, a flash of light spreads out in a spherical shell making a kind of hyper-cone in spacetime. That is not possible to visualise, so we'll stick to two dimensions of space to make drawing diagrams easier.

spacetime. They will be central to understanding black holes and the paradoxes they create. Let's zoom in on a particular event on our worldline to get more of a feel for the light cone and the relationships between neighbouring events in spacetime.

Christmas in spacetime

Let's imagine we've zoomed in on the region of spacetime in the vicinity of 'Christmas 1974' on our worldline. Your family are sitting around the TV arguing about whether to watch Bruce Forsyth and *The Generation Game* on BBC1 or Laurence Olivier in *Henry V* on BBC2. Presciently concerned about an incipient culture war, Granny springs to her feet and knocks a glass of Harvey's Bristol Cream onto the electric fire.* This causes the main fuse to blow, rendering the debate meaningless.

Figure 2.2 shows the spacetime region around 'Christmas 1974', drawn from the point of view of someone sitting in your house. Event A is 'Granny making contact with the glass of sherry' and event D is 'the fuse blows'. From this perspective, events A and D happen at very nearly the same place in space but at different times; D is in the future of A. The diagonal lines heading upwards and outwards from A trace out the future light cone of A. We've also drawn diagonal lines heading out into the past from A. These are known as the past light cone of A. All events in the shaded region inside the future light cone are timelike separated from A, which means that anyone who is present at A could also be present at any event inside the future light cone. All events inside the past light cone are also timelike separated from A. This means that anyone who is present at any event inside the past light cone could also be present at A. The expression for the interval between A and D is particularly simple: $(\Delta \tau)^2 = (\Delta t)^2$, where

---

* Both bars are on, and there is Vim under the sink.

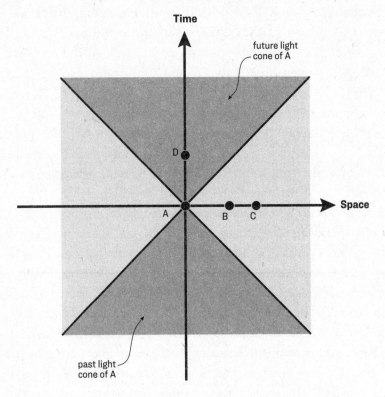

Figure 2.2. An event 'A' in spacetime and its neighbouring region. The diagonal lines are the lines traced out by beams of light that pass through A. They form the future and past light cones of event A.

Δt is the time difference between A and D as measured by a watch in your house.*

We've also marked two other events on the diagram, labelled B and C. From the house perspective, these happen at the same time

---

* Notice that, for pretty much all events we deal with in our everyday lives, $(\Delta \tau)^2 = (\Delta t)^2$ is approximately true. That's because the distances in space we are usually interested in range over a few metres or kilometres or even a few thousands of kilometres, and all of these are tiny when measured in light seconds. In everyday life Δx is very much smaller than 1 light second, and this is the reason why it feels to us as if time is universal.

as event A, but in a different place. Let's say they are an alarm clock going off at the other end of the street and a car starting its engine in the neighbouring town. The interval between A and B is $(\Delta \tau)^2 = -(\Delta x_{AB})^2$, and the interval between A and C is $(\Delta \tau)^2 = -(\Delta x_{AC})^2$. The interval is negative, which means that events B and C are spacelike separated from event A; Δx_{AB} and Δx_{AC} are distances that could be measured on a ruler.

Here is the key point. Event A caused event D (Granny knocked over a glass and that caused the fuse to blow). However, event A could not have caused events B and C. For that to happen, some influence would have to travel instantaneously from A to B and C because these things all happened at the same time. This delineating of causal relationships is why light cones are so important. Events inside each other's light cones can have a causal relationship because it is possible that some signal or influence could have travelled between them. Events outside each other's light cones cannot have a causal relationship. The interval therefore contains within it the notion of cause and effect. Certain events can cause others, and the light cones at each event tell us where in spacetime the dividing lines lie.

Let's now look at the same events in spacetime from two different perspectives. Figure 2.3 is a spacetime diagram constructed using measurements of distance and time made by an observer moving at constant speed past your house towards the car in the neighbouring town. As we have already discussed, such an observer will measure different times and different distances between events, but the intervals between events must remain the same because the interval is invariant. Nature doesn't care about your point of view, and the interval is a fundamental property of Nature. For this to be the case, something quite surprising happens. Events B and C happen *after* event A according to this observer.

Figure 2.4 shows a spacetime diagram constructed by an observer moving in the opposite direction at constant speed past

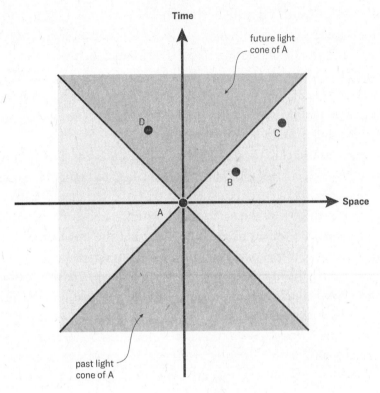

Figure 2.3. Events A, B, C and D as described in the text, as seen by an observer moving past event A at constant speed travelling from left to right on the diagram.

your house. This observer says that events B and C happen *before* event A.

At first sight, it seems that the spacetime picture has led to disaster. How can we countenance a theory that allows for the reversal in the time ordering of events? What if those events were your birth and death? Would someone be able to see you die before you were born?

The resolution to this apparent paradox can be seen by looking at the light cones. The light cones are in precisely the same place on all three diagrams, as they must be because all observers agree on the speed of light. Notice that although events B, C and D all

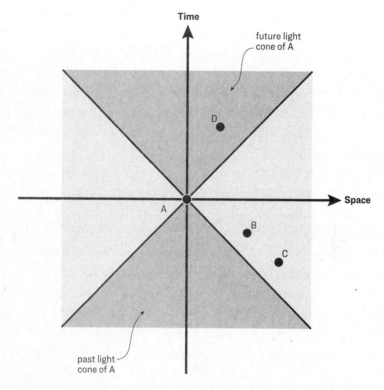

Figure 2.4. Events A, B, C and D as described in the text, as seen by an observer moving past event A at a constant speed travelling from right to left on the diagram.

move around on the spacetime diagram with respect to event A as we switch between different points of view, event D always remains inside the future light cone of A and events B and C always remain outside both the future and past light cones of A. To see that this must be the case, remember that the interval between two events is invariant: if the interval is timelike from one perspective, it's timelike from all perspectives. This means that events that can influence each other have their time-ordering preserved from all perspectives. Events that can't influence each other do not have their time-ordering preserved, but that doesn't matter because it does not mess with cause and effect. There is no contradiction if someone sees a house alarm going off or a car starting in the next town before or after Granny knocks over the glass, because these events could never have influenced each other – they are spacelike separated. There would of course be a contradiction if the lights fused before Granny knocked over the glass which caused them to fuse. But that can't happen for events A and D because D is always in A's future light cone, regardless of the point of view.

The future light cone of an event therefore tells us which regions of spacetime are accessible from that event and which regions are forbidden. Likewise, the past light cone of an event tells us which events in spacetime could have possibly had any influence on that event. If you look back at the worldline in Figure 2.1, you'll see that travelling to revisit moments in your past, the people and memories left behind, is impossible because we can never move from inside to outside the light cone at any event in our lives. To do so, we would have to travel faster than light. But the interval is invariant, so we can't do that. In a sense, our memories are out there, somewhere in spacetime, but we can never revisit them.

The picture of spacetime we've described above is that contained in Einstein's Special Theory of Relativity, first published in 1905. It describes a Universe without gravity, which is why we took the unconventional step of switching gravity off when discussing England versus Pakistan in Cape Town. Incorporating gravity into the spacetime picture is the concern of Einstein's General Theory of Relativity, published in 1915.

From special relativity to general relativity

The central idea of general relativity is that spacetime has a geometry that can be distorted. As we'll see, this corresponds to changing the rule for the interval between events. Matter and energy distort spacetime in their vicinity, and Albert Einstein

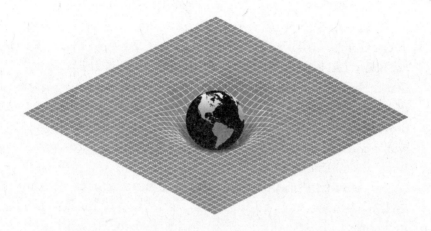

Figure 2.5. Schematic picture of the distortion of spacetime in the vicinity of the Earth.

worked out the equations that allow us to calculate how it is distorted. This is shown schematically in Figure 2.5. Objects like the International Space Station moving close to the Earth will be travelling through a region of distorted spacetime, and if we were Newtonians, we would interpret its motion as being due to a force deflecting it from a straight line and into orbit. But there is no force in Einstein's picture: gravity is to be understood purely as geometry.

An entirely reasonable first response to that last paragraph is to ask what on earth are we talking about. What does it mean to talk about distorting space and time? How should we picture distorted spacetime? Thus far, we have been drawing diagrams like Figure 2.1, with time pointing upwards and one or two directions in space represented horizontally. But our world isn't like that. We live in three spatial dimensions: forwards and backwards, left and right, up and down. Adding a fourth dimension – time – is very hard to picture.

To help us to comprehend the idea of spacetime, let's take a step back and imagine a two-dimensional world called flatland, populated by flat creatures.* The flatlanders can wander around, forwards and backwards, left and right, but never up or down. Their flat eyes can only see flat things on the flat surface and their flat brains can only comprehend flat things. Imagine the reception that renowned flat physicist Flat Albert would receive if he dared to say that space is really three-dimensional. 'There is another dimension, another direction we cannot point in,' he claims, and with his mathematics he would have no difficulty describing this three-dimensional world.

Suppose that Flat Albert is correct, and the flatlanders actually do live on the surface of a large messy table in an office, as shown in Figure 2.6. The third dimension is real; it's the direction upwards from the surface of the table, but the flatlanders can't see it. Their explorations haven't yet revealed that space comes to an end at the edge of the table, but they have discovered that there are impenetrable regions where they cannot go. They have to walk around the coffee cup and the lamp and books, and they are left to wonder at why the forbidden regions are sometimes circular, sometimes rectangular and occasionally some other less regular shape. Moreover, the land is covered in regions of light and dark that shift and change in shape and size.

How did Flat Albert deduce that there is an extra dimension in the world, based solely on observations made on the flat tabletop? 'It's all to do with those changing light and dark regions,' he says. 'I know what they are. They are shadows.'

Albert used mathematics to figure out that the shadows are two-dimensional projections of objects that live in three dimensions (coffee cups and books) and to deduce the three-dimensional shapes of the objects that cast them. It helps that the shadows change shape occasionally, which Albert correctly interprets as the higher dimensional source of brightness changing. From our

---

* We are inspired by Edwin Abbott's 1884 novel *Flatland: A Romance of Many Dimensions*. And possibly Mr Oizo's 'Flat Beat' featuring Flat Eric.

Figure 2.6. Flatland.

three-dimensional viewpoint, we immediately see that this is due to someone moving the table lamp.

Perhaps you can see the analogy. The interval – the thing that doesn't vary with point of view – lives in the four dimensions of spacetime. Distances in space and differences in time are mere shadows; they vary as we adopt different points of view in the three-dimensional world of our everyday experience. We can't picture something that lives in four dimensions, just as Flat Albert couldn't picture a coffee cup or a lamp or a book. But that didn't stop him lifting his gaze from the two-dimensional world of his experience to marvel at the true reality of three-dimensional space and the unchanging objects that live on the tabletop.

By pushing the flatland analogy a little further, we can also get a feel for how general relativity fits into this picture. The flatlanders may be inclined to assume that their tabletop is flat. If this were the case, parallel lines across flatland would never meet and the interior angles of triangles would add up to 180 degrees. We call such a flat geometry 'Euclidean'.

If the table is slightly warped, however, the flatlanders will discover small deviations from Euclid. Using precise measuring devices, they will be shocked to find triangles whose angles do not add up to exactly 180 degrees and parallel lines that converge or diverge from each other. It is in precisely this sense that we speak of space being curved or warped in Einstein's theory of gravity and it is what we are aiming to illustrate in Figure 2.5.

Flatland helps us to picture how space could have more dimensions than we directly experience, and it helps us to picture how space can be warped. By dropping down a dimension we can see the bigger picture of a two-dimensional space (the tabletop) embedded in a three-dimensional space (the room). We can't step outside of ourselves to see the bigger picture of four-dimensional spacetime because our imaginations are limited to picturing things in three dimensions or less. In that sense we are very much like flatlanders, fated to view the world in too few dimensions.

It's not easy to become comfortable with the idea of higher-dimensional spaces, but if it offers some comfort, professional physicists are no better at picturing four-dimensional spacetime than you are. When it comes to spacetime, we are all Flat Albert, peering at shadows. Fortunately, it is not necessary to try to visualise spacetime in all its four-dimensional grandeur. Often, we can drop a couple of dimensions in our mental picture and not lose anything important. We've already seen this in the spacetime diagrams we have used to explore the basics of special relativity, where (apart from Figure 2.1) we depicted only a single space dimension and the time dimension. Our understanding wouldn't have been enhanced by trying to draw two-dimensional space, although it might have made our diagrams look prettier. If we'd attempted to draw all three space dimensions plus the time dimension, we'd have run into problems.

We will rarely need to keep track of more than one space dimension in our study of black holes, because our focus will be on the distance from the black hole. We will be most interested in how the warping of the geometry influences the way cause and effect play out and that means keeping track of the light cones. Physicists have had a century to come up with a nice visual scheme for doing this, and the most widely used is named after Sir Roger Penrose. In the following two chapters, we'll introduce 'Penrose diagrams'. Armed with these beautiful maps of spacetime, we'll be ready to navigate beyond the horizon.

3

BRINGING INFINITY
TO A FINITE PLACE

Physicists often describe general relativity in aesthetic terms; it is the theory to which the word 'beautiful' is most often attached. 'Beautiful' implies an elegance and economy not easily visible in the mathematics, which is notoriously arcane. There is a well-known anecdote about Arthur Eddington who, when it was put to him that he was one of only three people in the world who understood Einstein's theory, paused for a moment and replied, 'I'm trying to think who the third person is.' Rather the adjective applies to the elegance and economy of the ideas that underlie the theory and to the beautiful idea that gravity *is* geometry. John Archibald Wheeler expressed this central dogma of general relativity in a single sentence: 'Spacetime tells matter how to move; matter tells spacetime how to curve.' The hard part of general relativity is to calculate how spacetime is curved, and for anything other than very simple arrangements of matter and energy, exact solutions to Einstein's equations are not easy to find. Black holes are one of the few cases in Nature for which we can precisely calculate the spacetime geometry, and once we have the geometry, we can represent it pictorially. The challenge is to find the most useful way of drawing the spacetime around a black hole on a flat sheet of paper. Flat paper is two-dimensional and spacetime is four-dimensional, which makes it difficult to draw (to say the least). If the spacetime is curved, that introduces an additional

Figure 3.1. *Hand with Reflecting Sphere* by M. C. Escher, 1935.

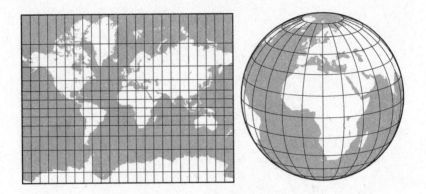

Figure 3.2. Mercator projection of the Earth's surface between 85°
South and 85° North.

headache and distortion is inevitable. The trick is to draw the
minimum number of dimensions necessary, which as we saw in
the previous chapter is often a single dimension of space plus the
time dimension, and to choose the distortion such that the
features we are interested in are rendered in a way that enhances
our understanding.

There is an image with which we are all familiar that introduces
distortion in a well-chosen way to represent a curved surface on a
flat piece of paper – a map of the surface of the spherical Earth.
Many ways of representing the surface have been devised but the
one of particular interest is shown in Figure 3.2. The Mercator
projection, introduced in 1569 by Gerard de Kremer,* is designed
specifically for navigation. Sailors care about compass bearings,
and presumably other stuff that's not relevant to this book, so the
Mercator projection is defined such that angles on the map at any
point are equal to compass bearings on the Earth's surface at that
point. This means that a navigator can draw a straight line on the
map between two places and the angle between the line and the

---

* De Kremer renamed himself Gerardus Mercator Rupelmundanus, Mercator being a
Latin translation of Kremer, which means merchant.

vertical will be the bearing from North that the ship must sail to travel between those places. The price to pay is that distances on the map are distorted, and the distortion increases with latitude. Greenland appears to be the same size as Africa on the Mercator projection, when in fact in area it is over 14 times smaller. The distortion becomes infinite at the poles, which cannot be represented on the map. The Mercator projection is an example of a 'conformal' projection, which means that angles and shapes are preserved at the expense of distances and areas.

The spacetime diagrams we drew in the last chapter extend infinitely in every direction: time carries on forever up and down, and space extends forever left and right. That's not necessarily a problem unless we are interested in depicting the physics of forever, but that's precisely what we would like to do if we want to visualise the spacetime in the vicinity of a black hole. If we are to develop an intuitive picture of a black hole, therefore, we'd like a way of bringing infinity to a finite place on the page. Roger Penrose found an elegant way of doing so.*

Penrose diagrams

Figure 3.3 shows the Penrose diagram for 'flat' spacetime. By flat spacetime, we mean a universe without gravity, which is the spacetime of special relativity we discussed in the last chapter. Flat spacetime is often referred to as Minkowski spacetime, after Hermann Minkowski who first introduced the idea of spacetime: 'The views of space and time which I wish to lay before you have sprung from the soil of experimental physics, and therein lies their strength. They are radical. Henceforth, space by itself, and time by itself, are doomed to fade away into mere shadows, and only a

---

* Sir Roger introduced his diagrammatic methods in the early 1960s and they were later used to great effect by Australian theorist Brandon Carter. Today, conformal spacetime diagrams are often referred to as Carter–Penrose diagrams.

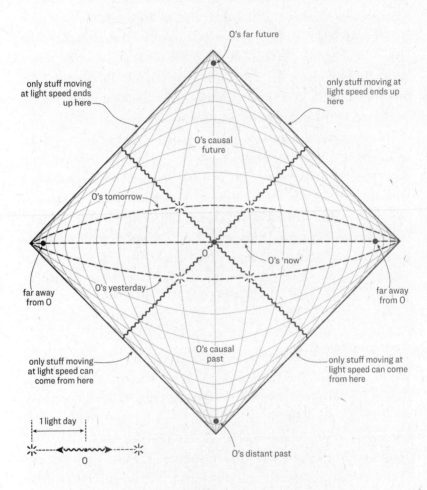

Figure 3.3. The Penrose diagram for flat spacetime in the simplified case of one space dimension.

kind of union of the two will preserve an independent reality.'* As we've discussed, matter and energy distort spacetime, and so in the presence of a planet, star or black hole, the geometry will change. But to warm up, we will focus first on simple, common or garden flat spacetime; a universe with nothing in it to distort the fabric.

The Penrose diagram for flat spacetime is a thing of beauty: all of space and all of time has been squeezed into a finite-sized diamond-shaped region. Every event in the history and future of an infinite, eternal universe is located somewhere on this diagram. The distortion is extreme, as it must be (we have captured infinity on a single page) but the distortion has been carefully chosen.† Just as for the Mercator projection, the Penrose diagram is a conformal projection; angles are preserved at the expense of distances on the page. This means that light rays always travel along 45-degree lines, and all the light cones are oriented vertically, just as they were in the spacetime diagrams of the last chapter. At any point, therefore, we can think of time as pointing vertically upwards. Since the light cones define the notion of past and future and tell us whether a particular event can influence another event, they are of utmost importance. If we are interested in how events in spacetime are related to each other in the vicinity of a black hole, a simple and intuitive rule for light cones is what we're after. The other beautiful feature of the Penrose diagram is that, as advertised, it brings infinity to a finite place on the page; not just one place, but five.

The diamond in Figure 3.3, representing an infinite, flat universe, is centred on a particular event, which we've labelled 'O'. There is nothing special about this central point. We could

---

*   Hermann Minkowski, in his 1908 address 'Raum und Zeit' to the Society of German Natural Scientists and Physicians.

†   Distortion in the sense used here has nothing to do with the distortion of spacetime due to matter. We are talking about the need to distort the representation to fit it on a sheet of paper.

choose any event in spacetime around which to centre the diagram, but of course we'll usually choose an event that we are interested in. It would be an eccentric choice to depict things we want to study in the most distorted regions out towards the corners of the diamond, but if we really want to then there's nothing stopping us. Remember that we are capturing all of space and all of time in this diagram, so every event that ever happened and ever will happen is depicted somewhere. 'All of space' is limited to one dimension – left and right – just to make things easier to draw. We could have drawn another spatial dimension, but it wouldn't add much to our understanding and would complicate the diagram. We've shown what adding another spatial dimension looks like in Box 3.2. The freedom to choose where to centre the diagram is also available for the Mercator projection. If we were interested in navigating around the polar regions, for example, we could choose Singapore as the 'pole'. The distortion would then be extreme around Singapore, but the map would be good for Greenland.

Since we can choose any event in spacetime around which to centre the diagram, let's go with an important one: your birth. For the purposes of this discussion, and only this discussion, the entire Universe – or at least this representation of it – revolves around you. The diagonal wavy lines that pass through O mark out the future and past light cones of O. We've labelled these two regions inside the light cones as 'O's causal future' and 'O's causal past'. Any event inside the future light cone can be reached from O and any event inside the past light cone can communicate with O. Since O is your birth, your worldline must snake around inside the future light cone.

The light cones can be thought of as the paths over spacetime of two pulses of light that started their journey in the distant past and happen to pass each other at event O before heading off into the far future. One light beam heads in from the left and another from the right. Remember that there is only a single dimension of space represented on this diagram. We've sketched this on a 'space-only' diagram at the bottom left of Figure 3.3. Two light beams travel towards and then past O from opposite directions, crossing at O. The flashes, marked on both diagrams, tell us the location of the light beams one day before they reach O and one day after they pass O.

We said that the light beams began their journey in the distant past and headed off into the far future. As we've drawn them, each beam originates from one of the bottom diagonal edges of the diamond and ends on the opposite top diagonal edge. The top edges are where all light beams end up if they fly through the Universe forever. You can see that any light beam emitted from any event on the diagram will end up there, because all light beams travel at an angle of 45 degrees. For this reason, the top edges are known as *future lightlike infinity*. Likewise, any light beam that was emitted in the infinite past will have begun its journey on one of the bottom edges. These edges are known as *past lightlike infinity*.*

Now let's focus on the grid lines on the figure. In the vicinity of O the grid looks like a sheet of graph paper, but closer to the edges the grid is increasingly distorted. This is the ultimate fish-eye lens effect: an infinite amount of space and time are squeezed into the diamond by shrinking spacetime by different amounts from place to place. The further away from the centre we go, the more shrinking is applied. On the Mercator projection of the Earth's surface, the *stretching* increases as we move away from the equator and an infinite *stretch* is applied at the poles, which is why we can't draw the polar regions. On the Penrose diagram, the *shrinking* increases as we head away from the centre of the diamond and an infinite *shrink* is applied at infinity, which is why we can represent an entire infinite universe within a diamond.

---

* Future lightlike infinity is often labelled $\mathcal{I}^+$ (pronounced 'scri plus') and past lightlike infinity is labelled $\mathcal{I}^-$ ('scri minus').

The distorted grid is a way to measure the coordinates of events in spacetime, just as the grid of latitude and longitude allows us to measure the coordinates of places on the Earth's surface. Looking at Figure 3.2, you can see how the distortion of the lines of latitude on the Mercator projection becomes more pronounced towards the poles, which affects the visual representation of the distance between points on the Earth's surface on the map. It is important to appreciate that the grid itself and the corresponding choice of coordinates is completely arbitrary. On Earth the choice is partly driven by the Earth's spin and the location of the geographical poles, but it's also historical. There is nothing about the geometry of a sphere that forces us to measure longitude relative to the meridian that passes through Greenwich Observatory in London.

Likewise, in spacetime any grid can be used although, just as for the Earth, some grids will be more useful than others. For example, non-rotating black holes are spherical and so when we are describing spacetime in the vicinity of a black hole we'll choose coordinate grids suited to a spherical geometry. It's worth emphasising, though, that the grids we choose don't even need to correspond to anybody's idea of space or time. It's just a grid, laid over spacetime such that we can label events. All that matters is that when we calculate the interval along a particular path using the grid, that distance will be an invariant quantity (which means it is independent of the grid choice). Analogously on Earth, the distance between London and New York doesn't depend on how we choose to define latitude and longitude.

That said, the grid on the Penrose diagram in Figure 3.3 does correspond to somebody's idea of space and time: yours. Let's focus on the moment of your birth, which we've identified as event O in the centre of the diamond. The horizontal dashed line passing though O represents all of space – 'now' – according to you. Every event on the 'now' line happened at the same time from your point of view at the moment you were born.

We now need to be clearer on what a 'point of view' is. Imagine the 'now' line is populated by a series of clocks which are all synchronised with the clock present at your birth. They are spaced at uniform intervals along the line; we might imagine them being connected by little rulers. Each clock is at rest relative to the clock at O. As time ticks, this line of clocks will march towards the top of the Penrose diagram into the future. The worldlines of these clocks are represented by the curving vertical lines on the diagram. The clock at O – the clock present at your birth and at rest relative to you when you were born – goes straight up the vertical line into the future. If you don't move, then the vertical line will also be your worldline. The other clocks also travel along what we will refer to as straight worldlines, even though they are curved on the Penrose diagram.

Let's say 'tomorrow' is exactly 24 hours after your birth. The whole line of clocks will have advanced up the diagram into the future and will now lie along the dashed line labelled 'O's tomorrow'.* Likewise, if 'yesterday' is exactly 24 hours before you were born, then all the clocks would be located along the line labelled 'O's yesterday'. All the curving horizontal lines on the diagram are therefore slices of space that are different moments in time from your point of view.† Only once we have been this careful can we say precisely what we mean by 'O's tomorrow' (the day after your birth). As this book unfolds, you will come to appreciate the significance of this apparent pedantry. If you stay still for your whole life relative to this ensemble of clocks, the horizontal lines are all your tomorrows, stacked up one after the other. Your tomorrows will end when your worldline ends at the event of your death, but the tomorrows on the Penrose diagram extend into the

---

* In three dimensions, the clocks populate all of space, not just a line.
† This picture can be extended to three dimensions of space. The clocks would form a three-dimensional lattice spanning all of space, and one could imagine a latticework of rulers connecting them together. In the terminology of relativity, such a latticework of clocks and rulers is known as an inertial reference frame.

infinite future. *Tomorrow, and tomorrow, and tomorrow, Creeps in this petty pace from day to day, To the last syllable of recorded time.**

We will now identify the remaining three of the five infinities on the diagram. Assuming that our imaginary clocks have always existed and always will exist, the worldline of every clock begins at the bottom vertex of the diamond and ends at the top vertex. If you recall from the previous chapter, anything other than light must move along a timelike worldline and could therefore carry a clock along with it.† Any (immortal) object that follows a timelike worldline will therefore begin at the bottom vertex of the diamond and end at the top vertex. The bottom vertex is known as *past timelike infinity* and the top vertex is known as *future timelike infinity*: 'the last syllable of recorded time'.

All the horizontal 'now' lines, representing infinite slices of space, begin and end at the left and rightmost vertices of the diamond. The spacetime distance between any two events on one of these slices is the distance as measured on a ruler between the events. The two remaining vertices of the diamond are thus identified with events that are an infinite distance in space from O and are known as *spacelike infinity*.

The Penrose diagram has brought infinity to a finite place. This ability to picture infinite space and eternity on a single diagram will be tremendously useful when we come to think about black holes. But first, let's have some fun with the Penrose diagram of flat spacetime in order to explore some of the famous consequences of Einstein's Special Theory of Relativity.

---

* Shakespeare's *Macbeth*.
† To be precise, we should say anything that is not massless.

Plate 1

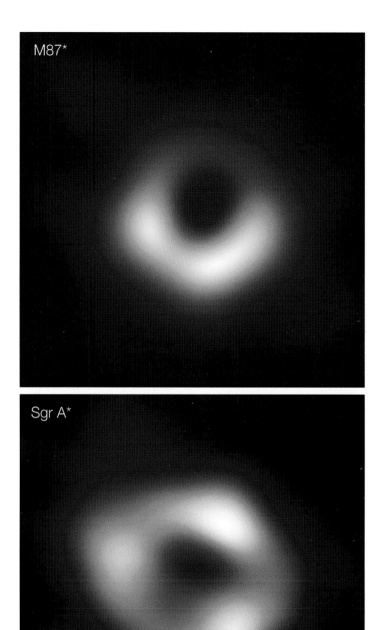

Figure 1.1. Top: The supermassive black hole at the centre of the galaxy M87. Bottom: Sagittarius A*, the black hole at the centre of our own galaxy. Both as imaged by the Event Horizon Telescope Collaboration. See page 5.

Plate 2

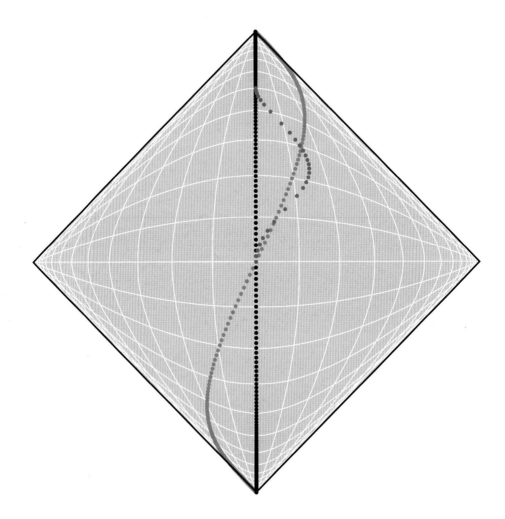

Figure 3.7. The Twin Paradox.
See page 64.

Plate 3

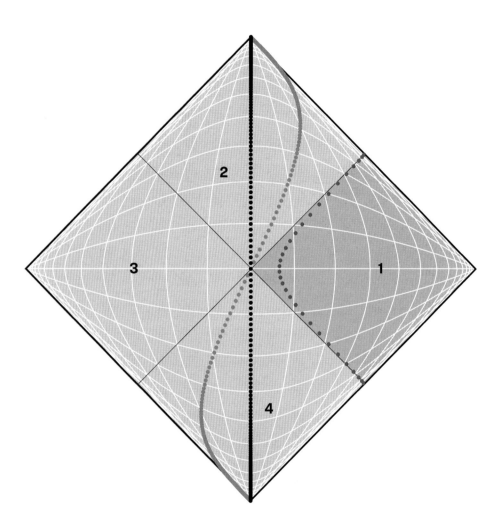

Figure 3.8. An immortal 'Rindler' observer
undergoing constant acceleration. See page 67.

Plate 4

Figure 4.1. The Montreal Biosphere, constructed for 'Expo 67'. See page 77.

Plate 5

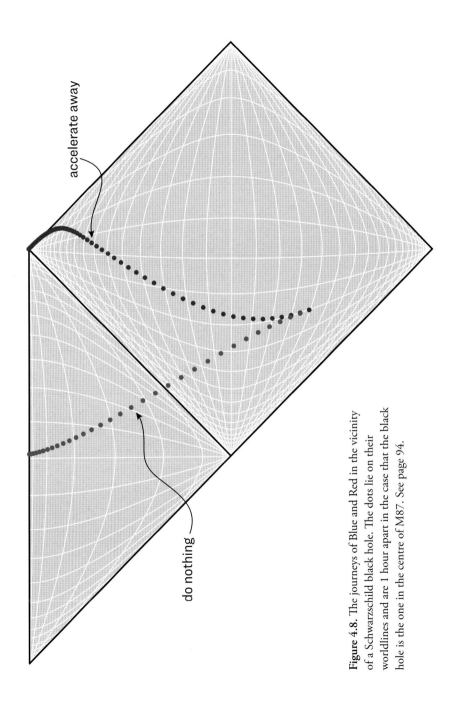

Figure 4.8. The journeys of Blue and Red in the vicinity of a Schwarzschild black hole. The dots lie on their worldlines and are 1 hour apart in the case that the black hole is the one in the centre of M87. See page 94.

Plate 6

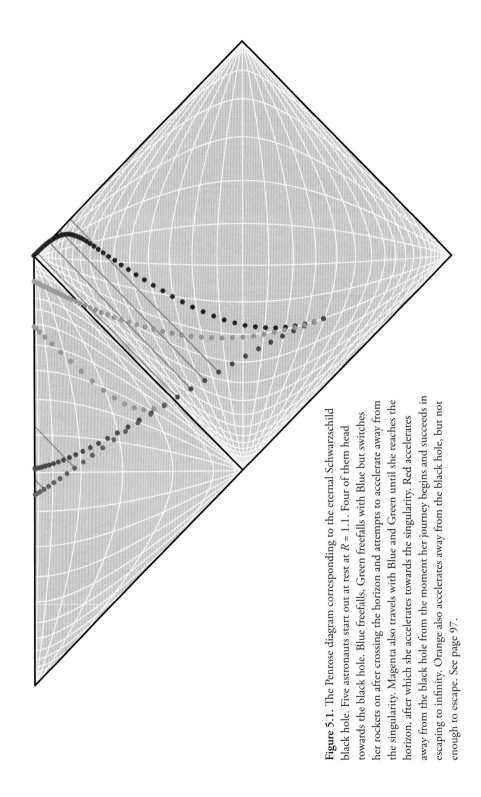

Figure 5.1. The Penrose diagram corresponding to the eternal Schwarzschild black hole. Five astronauts start out at rest at $R = 1.1$. Four of them head towards the black hole. Blue freefalls, Green freefalls with Blue but switches her rockets on after crossing the horizon and attempts to accelerate away from the singularity. Magenta also travels with Blue and Green until she reaches the horizon, after which she accelerates towards the singularity. Red accelerates away from the black hole from the moment her journey begins and succeeds in escaping to infinity. Orange also accelerates away from the black hole, but not enough to escape. See page 97.

Plate 7

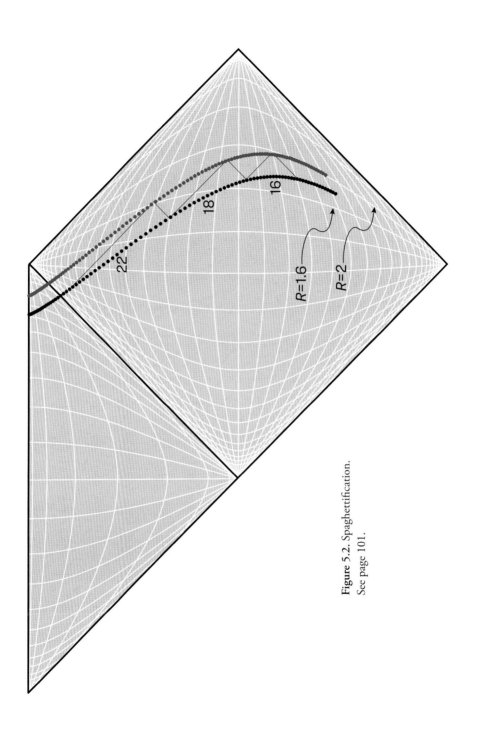

Figure 5.2. Spaghettification.
See page 101.

Plate 8

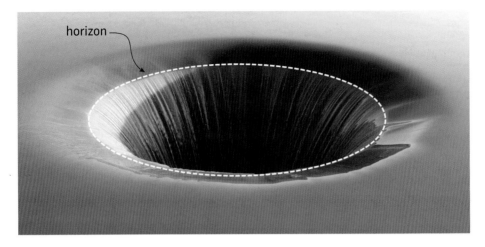

Figure 5.4. The river model of a black hole. See page 107.

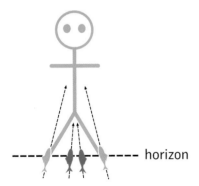

Figure 5.5. Photon-fish from Blue's feet enter Orange's eyes at a smaller angle than the photon-fish from Orange's own feet. Both enter the eyes at the same time, but Orange sees her own feet to be normal-sized and Blue to be distant and small. See page 110.

Plate 9

Figure 6.8. An astronaut (the red dot) falling into a Schwarzschild black hole. Time advances from the top left to bottom right. Notice how the wormhole grows and pinches off before the astronaut can reach the other side. See page 126.

Plate 10

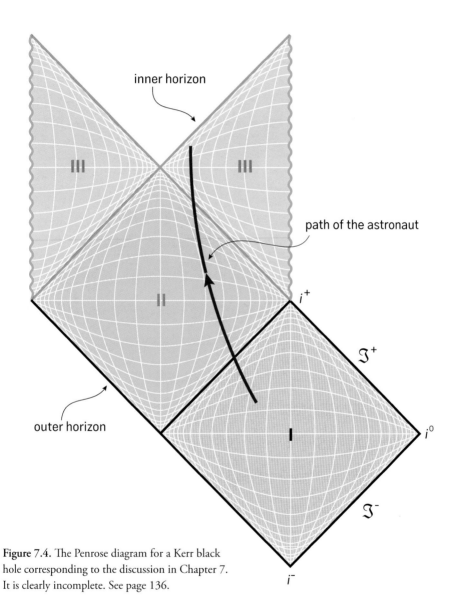

inner horizon

III

III

path of the astronaut

II

i^+

$\mathfrak{I}^+$

outer horizon

i^0

I

$\mathfrak{I}^-$

Figure 7.4. The Penrose diagram for a Kerr black hole corresponding to the discussion in Chapter 7. It is clearly incomplete. See page 136.

i^-

Plate 11

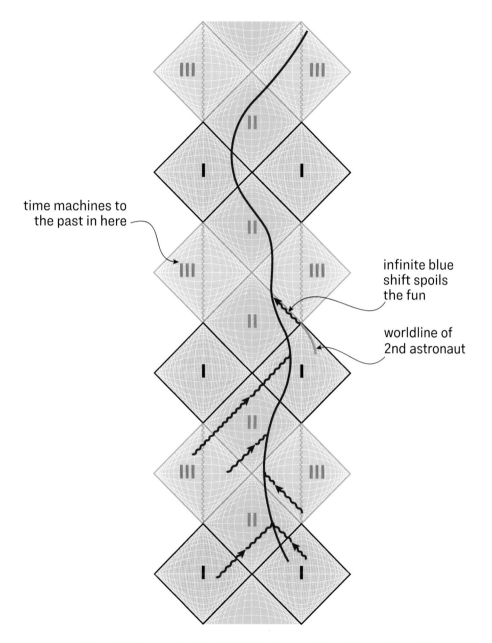

time machines to
the past in here

infinite blue
shift spoils
the fun

worldline of
2nd astronaut

Figure 7.5. The maximal Kerr black hole. The purple line is the worldline of our intrepid explorer. The part of her journey shown in Figure 7.4 is at the bottom. The wiggly purple lines represent possible paths of light rays. See page 136.

Plate 12

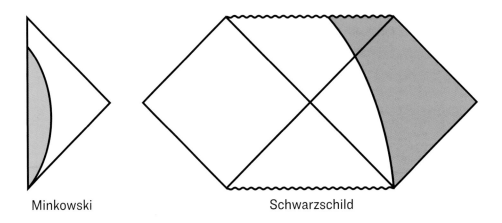

Minkowski   Schwarzschild

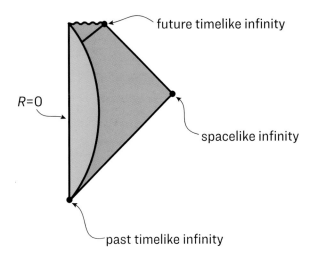

future timelike infinity

$R=0$

spacelike infinity

past timelike infinity

Figure 8.3. The Penrose diagram corresponding to a collapsing shell of matter (bottom). The inside of the shell is Minkowski spacetime (blue) and the outside of the shell is Schwarzschild spacetime (red). See page 157.

Plate 13

Figure 10.2. The temperature of a Schwarzschild black hole, written on Stephen Hawking's memorial stone in Westminster Abbey. See page 196.

Plate 14

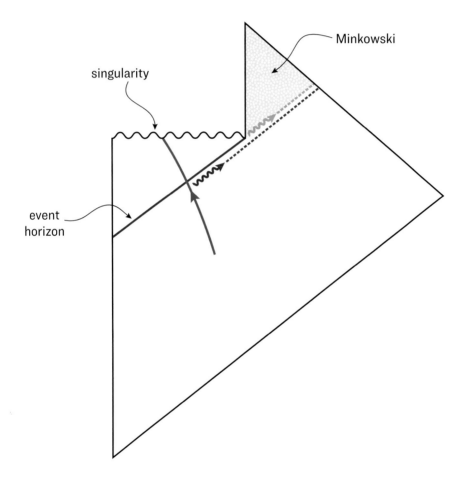

Figure 11.1. Penrose diagram for an evaporating black hole. The singularity disappears after the last Hawking particle is emitted (orange). The blue line is the wordline of a book thrown into the hole and in red is another Hawking particle. Both Hawking particles follow the dotted lines and end up at future lightlike infinity. See page 203.

Plate 15

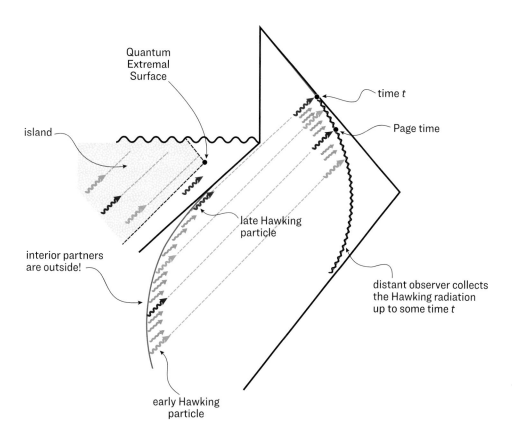

Figure 14.4. Part of the Penrose diagram corresponding to an evaporating black hole. The wiggly arrows denote Hawking particles and particles of the same colour are entangled partners (one is outside the event horizon and its partner is inside). The Quantum Extremal Surface corresponding to the radiation R is indicated, along with its island (the shaded region). Interior partners in the island should be considered to be part of R. See page 247.

Plate 16

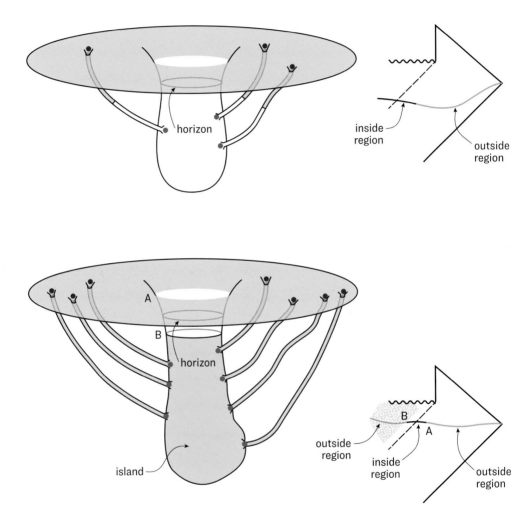

Figure 14.6. Illustrating how the island idea works. It vindicates both the ER = EPR idea and the Ryu–Takayanagi conjecture. For an old black hole (bottom), most of the 'inside' of the black hole is outside. In both pictures, the outside is shaded orange. See page 251.

The immortals

In the last chapter we were transported back to Christmas 1974 and family arguments around the television set, valves blazing. We also saw that events that happen simultaneously from one point of view do not happen simultaneously from a different point of view. More generally, observers moving relative to each other will disagree on the distance in space and the time difference between events, but they will always agree on the interval between them. Armed with our Penrose diagram, we can develop a more detailed picture of what's happening.

Let's consider two observers, Black and Grey, moving at constant speed relative to each other. Figure 3.4 shows a Penrose diagram with the worldlines of the two observers. They are immortals who have chosen to spend their infinite lives carrying out a visual demonstration for our benefit. Being immortal, their (timelike) worldlines begin at the bottom vertex of the diamond (past timelike infinity) and end at the top vertex (future timelike infinity). The immortals carry identical watches and have agreed to clap their hands every three hours. The dots along their world-lines mark the events in spacetime corresponding to their claps. The repetitive but illustrative cause embraced by the immortals for pedagogical benefit is more noble than that of Douglas Adams's character 'Wowbagger the Infinitely Prolonged' who decided to relieve the boredom of immortality by insulting every-body in the Universe in alphabetical order. He called Arthur Dent a 'complete kneebiter'. Wowbagger's worldline would begin inside the diamond, not at past timelike infinity, because he became immortal at a finite time in the past, allegedly in an accident involving a rubber band, a particle accelerator and a liquid lunch. His worldline would still end at future timelike infinity, leaving him plenty of time to complete his task.

The grid on the Penrose diagram is the same as that in Figure 3.3. It corresponds to a system of clocks and rulers at rest with

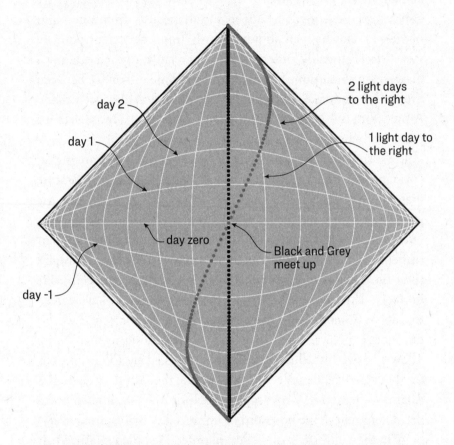

Figure 3.4. The trajectories of two observers moving over spacetime. Grey is moving at constant speed from left to right as determined by Black. The grid measures distances and times using a set of clocks and rulers at rest relative to Black.

respect to Black. Grey is moving at a constant speed from left to right relative to Black. Let's start by making sure we can appreciate that fact using the diagram. Black and Grey are at the same point in spacetime in the middle of the Penrose diagram, which means they fleetingly meet up there. Let's refer to the time when that happens as Day Zero. From her point of view, Black does not move through space, which means she travels along one of the vertically oriented lines in the grid. Since she started out in the middle of the diagram, she travels along the vertical grid line. If she'd started out from a point somewhere to the left or right then she would follow one of the curving vertical lines instead, but in both cases, she would not be moving relative to the grid. The curved appearance of a straight line is familiar to anyone who has been bored enough on a long-haul flight to stare at the map on the screen in the seat. Figure 3.5 shows the 'Great Circle' route from Buenos Aires to Beijing on a Mercator projection. This is a straight line on the curved surface of the Earth – the shortest distance between Buenos Aires and Beijing – but it looks curved because the map is a distorted projection of the surface of a sphere onto a flat sheet of paper.

Grey does move relative to the grid. Two days after passing Black we see he's travelled one light day away from her (according to Black's clocks and rulers, i.e. Black's grid). After a further two days of travel, Grey is two light days from Black, and so on. We can conclude that Grey is travelling at half the speed of light relative to Black.* It is worth checking that you understand the diagram well enough to see this before you read on.

Don't be confused by the fact that it looks like Black and Grey meet up in the distant past and the far future. They don't, because there is an infinite amount of space being squashed down at the

---

* Actually, Grey moves at 48.4 per cent the speed of light relative to Black, which you can just about see if you look really carefully at the figure. We chose 48.4 per cent for reasons that become clear in the next footnote.

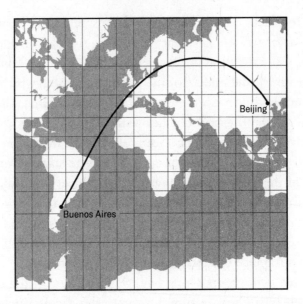

Figure 3.5. The Great Circle route from Buenos Aires to Beijing on a Mercator projection map. This is a straight line – the shortest distance between the two points on the Earth's surface.

top and bottom of the diamond (you can see all the grid lines bunch up there). The immortals only meet once, on Day Zero.

The next challenge is to use the diagram to see that Grey's watch runs slow compared to Black's. Look at Black's worldline. She claps her hands every three hours by her watch, which means she lays down eight dots every day along her worldline. Now look at Grey's worldline. He does the same, but according to Black's grid (i.e. Black's watch), he claps his hands only seven times per day. Crucially, this isn't some sort of optical illusion caused by the way we've drawn the diagram. Everything Grey does is slowed down as measured by Black, which means that Grey's whole life is running in slow motion from Black's point of view.*

---

* If you know a bit of special relativity (which you will if you have read our earlier book *Why does E = mc²?*) you might like to note that the factor 8/7 = 1.14 is equal to $1/\sqrt{(1 - v^2)}$ for $v = 0.484$.

BOX 3.1. The relativistic Doppler effect

At the risk of confusing matters, but for the sake of deepening understanding, notice that Black draws her conclusions by recording events using a grid, which we might think of as corresponding to a network of clocks and rulers at rest relative to her. Very importantly, she does not draw conclusions based on what she sees with her eyes. In fact, Black *sees* Grey living in fast-forward (the opposite of slow motion) before they meet on Day Zero, and in even slower motion when Grey has passed her. It's possible to work this out from Figure 3.4 and it is well worth the effort if you fancy a challenge.

Here's how it works out. Since light moves on 45-degree lines and since we see things using light, it follows that on day minus one (the day before Day Zero), Black sees Grey when he is at day minus two. During the next 24 hours, in the period leading up to their brief encounter at Day Zero, Black lays down the usual eight dots while Grey lays down 14 dots, which means Black *sees* Grey clap faster. After Grey passes by, things flip around and Black sees Grey clap slower; she *sees* Grey clap just under five times per day. We'll leave that for you to work out by counting dots and thinking about 45-degree light beams.

The point is that, in relativity theory, it is very important to say exactly how time differences are being determined. *Seeing* things (with instruments like eyes) can be very different from measuring the passage of time using a network of clocks and rulers. The effect we just discussed is known as the relativistic Doppler effect and it is sensitive to the location of the light detector (i.e. where the eyes are). That's why Grey went from fast-forward to ultra-slow-motion as he passed Black. There is a more familiar and similar effect for sound (also called the

> Doppler effect) in which we hear a change in the pitch
> of a siren when an ambulance drives past. The lesson is
> that we need to be careful about using the word 'see' in
> relativity.

Now let's change point of view and consider everything from Grey's perspective. In Figure 3.6 we've changed the grid such that it now represents measurements made using clocks and rulers at rest with respect to Grey. Relativity is so-called because of this relative aspect of motion – who is at rest and who is moving is just a point of view. Now Grey doesn't move, which means his world-line snakes along a grid line. As before, Grey and Black clap their hands once every three hours according to their individual watches, but now it's Black that claps only seven times a day. Grey concludes that Black is living her life in slow motion; their roles have been entirely reversed.

At this point you are well within your rights to exclaim loudly that this is nonsense. How can Grey age more slowly than Black according to Black's clocks while Black ages more slowly than Grey according to Grey's clocks? This sounds impossible, but surprisingly there is no contradiction. The 'problem' arises because, following in the footsteps of Newton,* we are fixating on the concepts of universal time and space. Instead, we need to rewire our brains and focus on the worldlines – the paths traced out over spacetime by the immortals – and the grids of rulers and clocks they erect to describe the world. Black's grid, shown in Figure 3.4 is different to Grey's grid, shown in Figure 3.6. The grid lines that run roughly horizontal across the Penrose diagrams represent all of space 'now' for each immortal. The grid lines that run vertically represent all of time. But the grids are not the same. Grey's space is a mixture of Black's space and time, and vice versa. It's

---

* Admittedly difficult when he himself is standing on the shoulders of giants.

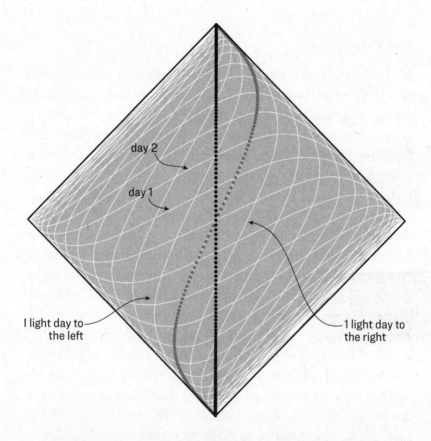

Figure 3.6. The trajectories of two observers moving over spacetime. Grey is moving at constant speed from left to right as determined by Black. The grid now represents a set of clocks and rulers at rest relative to Grey.

hard to accept that the delineation between space and time is subjective, because our personal experience is that they are fundamentally different things that cannot be mixed. But this is not true. The separation between them is personal; it depends on our point of view.

The Twin Paradox

All very well, you may say, but what happens if the immortals decide to meet up again in the future? Then we really would appear to have a paradox, because we'll be able to tell who has aged more. That's a direct observation of reality, and we can't then have it both ways. Indeed, we can't. This apparent paradox is sometimes known as the Twin Paradox.

To see why the Twin Paradox isn't a paradox, let's introduce a third immortal called Pink. We've added her worldline onto the Penrose diagram in Figure 3.7. We now have a triplet paradox. Our three immortals meet up for a fleeting moment at Day Zero. Grey zooms past Black at half the speed of light, exactly as before, and Pink uses her spaceship to fly along a path that allows her to meet up with Grey and Black again in the future. Let's look at Pink's worldline to work out what she's doing. After Day Zero, Pink accelerates away from Black, moving slowly at first, which is why their worldlines almost overlap for two handclaps. She then begins to speed up and catch up with Grey. When Pink and Grey meet (at the end of Day One), we can ask a question to which we must receive a definitive answer: who is older, Pink or Grey? Counting the dots along their respective worldlines informs us that Pink clapped her hands six times, while Grey clapped his hands seven times: Grey is older.

Though it is very counter-intuitive, the idea that Grey ages more than Pink is simple to appreciate if you recall from the previous chapter that the length of a timelike worldline is the time

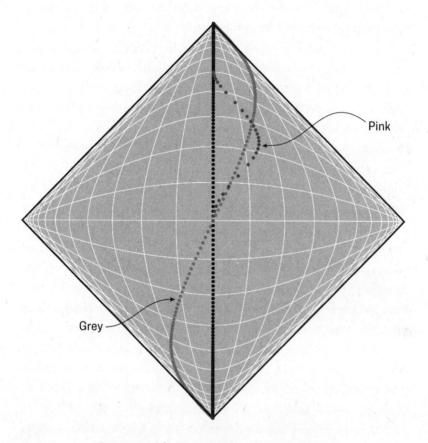

Figure 3.7. The Twin Paradox. Also on plate 2.

measured by a watch carried along that worldline. With that single idea, it's easy to see that Pink and Grey age at different rates because Pink's worldline is different from Grey's worldline between their meetings. What's not obvious without calculating is whose worldline is longer. You can see that by counting handclaps on the diagram; we did the calculation for you using the space-time interval equation from Chapter 2.

Let's continue following Pink's journey. After around a day, she's travelling very close to the speed of light, which you can see from the Penrose diagram because her worldline is almost at an angle of 45 degrees. She then swings her spaceship around and fires her rockets to decelerate, eventually reversing direction. She meets Grey again, and the two immortals can compare their ages once again. Counting the handclaps, Pink has aged 17 x 3 = 51 hours and Grey 20 x 3 = 60 hours since they first met at Day Zero. Finally, Pink returns back to Black. 28 x 3 = 84 hours have passed on her watch, and 120 have passed on Black's watch (you can see this without counting dots because this rendezvous occurs after five days using Black's grid, which is shown on the figure). This is nothing less than time travel into the future. Black has aged more than Pink when they meet up. Fascinatingly, there is no limit on how far into the future one can travel with access to a fast enough spaceship. The Andromeda galaxy is 2.5 million light years away from Earth. If Pink had access to a spaceship that could travel at 99.9999999999 per cent of the speed of light, it would take her 18 years to make the round trip to Andromeda. She would, however, return to Earth 5 million years in the future.

There is a general principle at work here known as the Principle of Maximal Ageing. Black and Grey will age more than anyone else who sets off from Day Zero and takes any route over space-time before returning to meet up with them again. The thing that is special about Black and Grey is that they never turn a rocket motor on to accelerate or decelerate. We call Black and Grey's routes between events 'straight lines' over spacetime because they don't accelerate.*

Horizons

For our final foray into flat spacetime, we'll explore acceleration, and in doing so follow in Einstein's footsteps on the road to general relativity. In Figure 3.8, the purple dotted line corresponds to an immortal who starts out in the distant past travelling from right to left at close to the speed of light. We've named this immortal 'Rindler', after the physicist Wolfgang Rindler, who first introduced the term 'Event Horizon'. Rindler steadily decelerates until he reaches his closest approach to Black at Day Zero. From the diagram, you should be able to see he is momentarily stationary relative to Black at a distance of just over half a light day. His constant acceleration then takes him away again, out towards infinity, moving all the time ever closer to the speed of light. For the first half of the journey, the rockets are slowing him down and for the second half they are speeding him back up again relative to Black. Rindler is always accelerating at the same rate, as measured by accelerometers onboard his spaceship, although he won't need instruments to tell him that he's accelerating. He will feel the constant acceleration as a force pushing him into his seat. He won't be 'weightless' inside his spacecraft. Hold that thought, because it's going to be very important.

This trajectory of a constantly accelerating observer is known as a Rindler trajectory. Notice that although Rindler accelerates

---

* Incidentally, you might like to note that the straight-line path between events in spacetime is longer than any non-straight line. This is because the geometry of spacetime is not the geometry of Euclid. If it were, the interval would be given by Pythagoras' theorem: $(\Delta\tau)^2 = (\Delta t)^2 + (\Delta x)^2$. But the interval contains a minus sign: $(\Delta\tau)^2 = (\Delta t)^2 - (\Delta x)^2$, and that makes all the difference. The geometry of flat spacetime is what mathematicians call hyperbolic geometry.

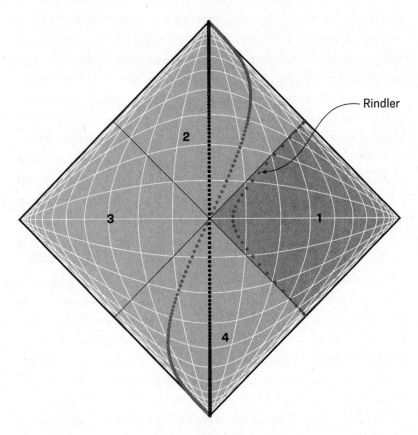

Figure 3.8. An immortal 'Rindler' observer undergoing constant acceleration. Also on plate 3.

forever, his worldline never quite makes it to 45 degrees on the Penrose diagram. That is because, no matter how long he accelerates for, he cannot travel faster than the speed of light. The most striking thing about Rindler's trajectory is that he always remains inside the shaded right-hand region of the Penrose diagram we've labelled '1'. He can see anything that happens in this region at some point during his journey. By 'see' we do mean 'see' in the sense that light can travel from any event inside region 1 and reach his eyes. To confirm this, pick any point in region 1 and check that 45-degree light beams emitted from that point will intersect Rindler's trajectory. Likewise, Rindler could have sent a signal to any event in this region of spacetime at some point during his journey.*

Can you see that Rindler cannot receive signals from regions 2 and 3? This is because there is no way for anything travelling less than or equal to the speed of light to get from those regions into region 1. We say that regions 2 and 3 are beyond Rindler's horizon. In fact, region 3 is particularly isolated because it is also impossible for anyone in region 3 to receive signals from Rindler. These two regions are completely causally disconnected. Region 4 is different; Rindler could receive signals from this region, but he couldn't send signals into it. Rindler's situation is very different to that of Black and Grey, for whom the entirety of spacetime is causally accessible. Rindler lives in a smaller Universe than Black and Grey. By virtue of his acceleration, he has cut himself off from some regions of spacetime. The 45-degree boundary lines to his region are generically referred to as 'horizons' because information cannot flow both ways across them.

Previously we encountered horizons in the context of gravity and black holes. Now we see that they also appear for accelerating observers. Is there a conceptual connection between acceleration

---

* Another way to see this is to note that, because Rindler starts at the bottom of region 1, all of region 1 lies within his future light cone.

and gravity? Indeed there is, and when Einstein first realised it, he called it the happiest thought of his life.

The happiest thought

We've all seen pictures of astronauts aboard the International Space Station. They float. If an astronaut lets go of a screwdriver, it floats next to them. Even globules of water float around undisturbed as mesmerising, gently oscillating bubbles of liquid. Why? The astronauts, the screwdriver, the water and indeed the International Space Station itself have not escaped Earth's gravitational pull. They are only 400 kilometres or so above the surface: just ten times the altitude of a commercial aircraft. If you were to jump out of an aircraft, you would be unwise to assume you've escaped gravity and not deploy your parachute. The space station is falling towards the Earth in just the same way as you would if you jumped out of an aircraft, but it's travelling fast enough relative to the surface of the Earth – around 8 kilometres per second – to continually miss the ground. It can continue to orbit in this way with very little intervention from rockets because there is very little air resistance at an altitude of 400 kilometres. We say the space station is in freefall around the Earth; forever falling towards the ground but never reaching it. The crucial point is that freefall is locally indistinguishable from floating freely in deep space, far away from any stars or planets; that is to say, if the astronauts had no windows and could not look outside to see the Earth below, they would be unable to do any experiment or make any observation to inform them that they are in the gravitational field of a planet. This is the reason every object floats undisturbed in the space station; there is no force to disturb them, and this is the idea that Einstein famously described as the 'glücklichste Gedanke meines Lebens', the happiest thought of my life. It immediately suggests that there is something interesting about the force of gravity, because gravity can be removed by falling. Likewise, its

Figure 3.9. Rindler's spacecraft.

effects can be simulated by accelerating. Acceleration is locally indistinguishable from gravity, and vice versa.* That very important idea is known as the Equivalence Principle.

Imagine that Rindler accelerates at 1g.† Inside his spacecraft, Rindler's experience would be precisely the same as if he were sitting comfortably in an armchair or wandering around his cabin on the surface of the Earth. If he had no windows, there is no experiment or observation he could perform to tell him otherwise. If he reduced the power of his rocket and dropped his acceleration

---

* We say freefall is *locally* indistinguishable because the Earth's gravitational field is not uniform, and this is detectable over large-enough distances. For example, objects fall towards the centre of the Earth, and this means that two objects which begin falling parallel to each other at some height will get closer together as they head towards the ground. This is called a tidal effect. We'll meet these effects, which in the context of black holes lead to spaghettification, later in the book.

† 1g is an acceleration equal to that of an object falling in the vicinity of the Earth's surface, i.e. 9.8 metres per second squared.

down to around 0.3g, he might imagine he was sitting on the surface of Mars. No other force in Nature behaves like this. It's not possible to remove the force between electrically charged objects by accelerating or moving around. And yet this is possible for gravity. This is the clue that led Einstein to formulate his theory of gravity purely in terms of the geometry of spacetime. Gravity as geometry. Let's explore that remarkable idea.

BOX 3.2. Extending the Penrose diagram to two space dimensions

We have been working in a world of one space dimension where our observers can only move along a line. Much of relativity theory can be understood without needing to invoke the other two space dimensions we move around in, just as we can appreciate much of Newton's mechanics by considering things moving along a straight line. But we should talk a little about those other two space dimensions. The left-hand picture of Figure 3.10 shows the Penrose diagram of flat spacetime in 2+1 dimensions (this is the standard notation for 2 space dimensions and 1 time dimension). It looks like two cones, glued base-to-base. At any given moment in time, we can draw a 'now' surface, rather than the 'now' lines we've been thinking about in our 1+1 dimensional Penrose diagrams throughout this chapter. The 'now' surface at the junction of the two cones (Time Zero) is a flat disk, and as we move forwards in time the Now surfaces become distorted into domes just as our lines were distorted into curves.

The right-hand diagram in Figure 3.10 is more closely related to the 1+1 dimensional diagrams we have been drawing throughout this chapter. Our diamonds are obtained by reflecting the triangle about the vertical dashed line. The complete 2+1 diagram is obtained by sweeping the

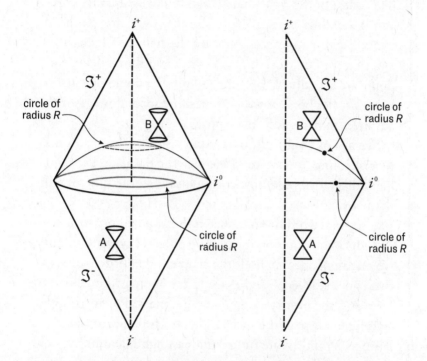

Figure 3.10. Extending the Penrose diagram to two space dimensions. The diagram on the left is obtained by rotating the one on the right about the vertical dashed line.

triangle around the vertical dashed line. The left-hand diagram is a complete representation of every spacetime point in 2+1 dimensions. The right-hand diagram loses some of the information because entire circles are drawn as single dots.

It's only in the 2+1 diagram that light cones actually look like cones. In the right-hand diagram the light cones at A and B look like 'crosses'. Note also that the dome on the left-hand diagram appears as a line on the right-hand diagram, and that the circle on the 'now' sheet and the circle on the dome are actually of the same radius. The dome-one just looks smaller because we are squashing space down as time advances to make it fit into the diagram.

We live out our lives in 3+1 dimensions. We can't draw that of course, but we can imagine. The points in the right-hand diagram would now correspond to entire spheres rather than circles.

One final word: we have taken the opportunity to introduce the notation for the five different types of infinity we mentioned throughout the chapter. The vertices of the two cones labelled i^+ and i^- are future and past timelike infinity – the ultimate origin and destination of anything travelling less than the speed of light. The cone's surfaces labelled $\mathfrak{I}^+$ and $\mathfrak{I}^-$ are future and past lightlike infinity, accessible only to light beams or anything else that can travel at the speed of light. The circle at the junction of the two cones marked i^0 marks out spacelike infinity – the infinitely distant regions of space at any moment in time.

4

WARPING SPACETIME

Less than five months before his death in 1916, while serving in the German army calculating the trajectories of artillery shells on the eastern front, the eminent astrophysicist Karl Schwarzschild discovered the first exact solution to the equations of Einstein's General Theory of Relativity. Schwarzschild's achievement was no less remarkable for the fact that he derived the solution and sent it to Einstein just a few weeks after the theory had been published. Einstein was impressed, writing in return, 'I have read your paper with the utmost interest. I had not expected that one could formulate the exact solution of the problem in such a simple way.' Schwarzschild had found the equation describing, to very high accuracy, the geometry of spacetime around a star. Recall John Wheeler's maxim: 'Spacetime tells matter how to move; matter tells spacetime how to curve.' Schwarzschild's solution describes the curve of spacetime, and it's then a reasonably straightforward task to work out how things move over it. Today, Schwarzschild's solution is one of the first things taught in an undergraduate course on general relativity and, in most circumstances, it corresponds to a tiny improvement over the simpler Newtonian predictions for planetary orbits. But not in all circumstances, because the Schwarzschild solution, unbeknown to him and to Einstein in 1916, also describes black holes.

What does Schwarzschild's solution to Einstein's equations look like? We caught a glimpse of the answer when we thought about the warped tabletop of flatland. Flat Albert and his flat mathematician friends discovered that the geometry of Euclid no longer applies because the table's surface is warped. The angles of triangles do not quite add up to 180 degrees and distances between points will not be described by the familiar form of Pythagoras' theorem. If Flat Albert wanted to calculate the distance between two places in warped flatland, he'd have to find a way of representing the warping mathematically.

To warm up, let's forget spacetime for the moment and return to Earth. The Earth's surface is curved, and this means we can't simply use Pythagoras' theorem to calculate the distance between widely spaced cities such as Buenos Aires and Beijing. We can easily appreciate this because we are three-dimensional beings, and we know what a sphere looks like. For example, if we were to procure a 19,267-kilometre-long ruler and place it in downtown Buenos Aires, its tip would not land in Beijing. The reason is that the ruler is flat and the Earth isn't. The ruler would stick out into the third dimension – its tip would end up off the surface, out in space. We could, however, imagine purchasing 20 million one-metre rulers and laying them end to end along the great circle route between the two cities. The rulers could be made to follow the curved contour of the Earth's surface (in reality we'd have to tunnel through any mountains we encountered to keep our chain of rulers at sea level, so we are imagining a perfectly smooth spherical Earth). Give or take a few metres, we could measure the distance over the curved surface of the Earth this way. If we wanted to do better, we could make each ruler one centimetre long or even smaller. Smaller rulers have the advantage that they better track the curving surface of the Earth.

The idea that we can build up a curved surface using lots of little flat pieces is nicely illustrated by the Montreal Biosphere, designed by Buckminster Fuller for the 1967 World's Fair. When viewed from afar, the biosphere is a perfect-looking sphere, but close-up we can see it's made up of lots of small flat triangles, 'sewn together' but slightly tilted with respect to each other. Any shape could be constructed in this way; the geometry is determined by how the flat pieces are assembled.

In general relativity, curved spacetime can be built up in the same way. Lots of little pieces of flat spacetime can be sewn together to make curved spacetime, and Schwarzschild's solution to Einstein's equations describes how they are sewn together in the vicinity of a star. The distance in spacetime between events on each little flat piece could be determined by an arrangement of clocks and rulers, i.e. the interval $(\Delta\tau)^2 = (\Delta t)^2 - (\Delta x)^2$. We can get a very accurate description by choosing the spacetime patches to be sufficiently small that the flat space formula for the interval is a good approximation over each patch. This is just like saying the distance between two points on one of the little triangles on Buckminster Fuller's biosphere can be determined using Pythagoras' theorem, despite the fact that the distance between

Figure 4.1. The Montreal Biosphere, constructed for 'Expo 67'.
Also on plate 4.

two points separated by many little triangles requires a more difficult calculation because the surface is curved.

We've sketched this patchwork view of spacetime in Figure 4.2. If we were five-dimensional beings with an innate sense of hyperbolic geometry, we would be able to visualise how the little pieces sew together to make a 'surface' curved into a fifth dimension. Good luck with that, but the basic idea is quite simple. We are to think of curved spacetime as being tiled by lots of little flat pieces of spacetime, each slightly tilted with respect to their neighbours and adorned with their own grids of clocks and rulers. The challenge in general relativity is to specify how the pieces are sewn together. If we know that, we can calculate the interval between widely separated events on the curved surface by adding up all the intervals on each patch, just as we laid down the little rulers to measure the distance between Buenos Aires and Beijing.

The idea that curved spacetime is well approximated by flat spacetime over sufficiently small distances and intervals of time, just as the Earth is flat over sufficiently small distances, is precisely what Einstein had in mind when he had his happiest thought:

'At that moment there came to me the happiest thought of my life … for an observer falling freely from the roof of a house no gravitational field exists during his fall – at least not in his immediate vicinity. That is, if the observer releases any objects, they remain in a state of rest or uniform motion relative to him, respectively, independent of their unique chemical or physical nature. Therefore, the observer is entitled to interpret his state as that of rest.'

This quote is wonderful because it illuminates Einstein's thinking. He didn't think mathematically, at least initially. He thought in simple pictures and asked simple questions. What does the fact that gravity can be removed by falling tell me? If gravity can't be detected in a freely falling observer's immediate vicinity, spacetime must be flat in their immediate vicinity. Don't get confused about what it feels like to actually fall off a roof by the way – we're

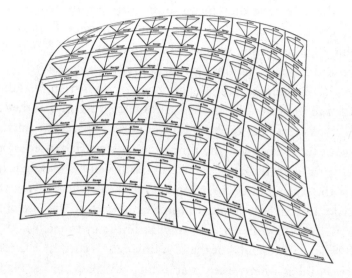

Figure 4.2. Building up spacetime by sewing together a patchwork of tiny regions each of which is flat.

considering an idealised fall in a vacuum and ignoring air resistance. Simplify the problem down to its essence. All cows are spherical to a theoretical physicist, which is why they are clear thinkers but shit farmers. Gravity is a strange force because it can be removed by falling. Einstein's genius was to see the connection between this idea and a geometric picture of gravity as curved spacetime. Gravity appears not because there is a fundamental force of attraction between things, as we learn at school, but because small patches of spacetime are tilted relative to their neighbours in the vicinity of massive objects.

If there is no force of gravity, why does a person fall off a roof and hit the ground, or the Moon orbit the Earth? The answer is that the person and the Moon are both following straight lines over curved spacetime. We can be more specific if we recall the Twin Paradox in the previous chapter. There, we encountered the Principle of Maximum Ageing. An astronaut that does not accelerate takes a path over spacetime between two events that maximises the time they measure on their wristwatch between those events. In general relativity, the Principle of Maximum Ageing is placed centre stage as a fundamental law of Nature that determines a freely falling object's worldline over curved spacetime. As Einstein says in his quote, an observer in freefall 'is entitled to interpret his state as that of rest'. This implies that the path a freely falling object takes over curved spacetime must be the path that maximises the time on a wristwatch carried by the object. On each little flat patch, this path will be a straight line across the patch, but in curved spacetime the patches sew together to make a curve. The result is entirely analogous to the case of the little flat rulers on Earth. The straight lines must match up with each other end-to-end, but the resulting path is curved. The result in spacetime is what we see as an orbit – the paths of the planets around the Sun. Or, for that matter, the fall as someone slips unfortunately off a roof. In a way, the path the unfortunate faller takes on their way to the ground is entirely logical – they are maximising the time they have left.

Schwarzschild's solution for the curvature of spacetime, when paired with the Principle of Maximum Ageing, is all we need to calculate the worldlines of anything falling in the vicinity of a planet, star or black hole.

The opportunity to acquire a deeper understanding of general relativity and Schwarzschild's solution lies within our grasp, and it would be a shame not to go all the way when we've come this far. So, over the next few pages there is a little more mathematics than in the rest of the book. There is nothing much more complicated than Pythagoras' theorem, but if you really don't like mathematics then don't worry; normal diagrammatic service will be resumed shortly.

The metric: calculating distances on curved surfaces

By 1908, Einstein had the basic idea of gravity as curved space-time, but it took him a further seven years to achieve its mathematical realisation in the form of general relativity. His challenge was to find a way of calculating the distance between two events if spacetime is curved. When asked why it took so long, here is what he said: 'The main reason lies in the fact that it is not so easy to free oneself from the idea that coordinates must have an immediate metrical meaning.'[17]

To understand what Einstein meant, we'll leave spacetime for a moment and return to two-dimensional Euclidean geometry. Let's choose two points A and B and draw a straight line between them, as shown in the left-hand picture in Figure 4.3. The line will have a length that we could measure with a ruler. Call that length Δz. The line of length Δz is also the hypotenuse of a right-angled triangle with sides of length Δx and Δy. Pythagoras' theorem relates these three lengths:

$$(\Delta z)^2 = (\Delta x)^2 + (\Delta y)^2$$

In this equation, all the quantities are distances that can be measured by rulers. They also happen to be the difference in coordinates using the grid that you can see in the background. Specifically, A is at $x = 3$ and $y = 2$, which we write (3,2), and B is at (9,7), so $\Delta x = 9 - 3 = 6$ and $\Delta y = 7 - 2 = 5$. Using Pythagoras gives us $\Delta z = \sqrt{61}$.

Now look at the right-hand picture in Figure 4.3. It is the same pair of points, A and B, but now with a different grid. The new grid is perfectly good for labelling points: A is at (5,3) and B is at (7,9). But the differences in these coordinates cannot be used in Pythagoras' theorem. This is a nice illustration of the arbitrariness of grids. Any grid will do for labelling points, but some grids are

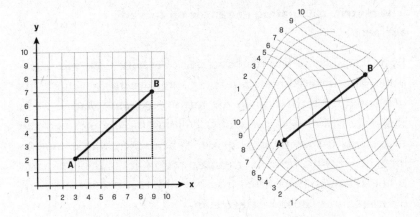

Figure 4.3. Left: The distance between A and B is related to the coordinates of A and B via Pythagoras' theorem. Right: The same two points, A and B, can be located using a wavy grid but the distance between them is not related to the coordinates via Pythagoras' theorem.

more useful than others. Here, the square grid makes it easier to calculate the length from A to B. On a flat surface, we can always choose a rectangular grid to make life easier, but on a curved surface, it is impossible to pick a single grid such that the Pythagorean equation works for the distance between every pair of points. The distinction between the coordinate grid as a mesh for labelling points and as a device to help calculate distances between points is what Einstein was referring to when he said it is 'not so easy to free oneself from the idea that coordinates must have an immediate metrical meaning'. We need to follow his words of warning and not get too attached to the grids we lay down over spacetime.

What is true is that there is always a way to compute distances using any grid. It's just that the formula is not the one due to Pythagoras. If X and Y are the coordinates labelling the wavy grid on the right of Figure 4.3 then, for any two points that are sufficiently close together the distance between them can always be written:

$$(dz)^2 = a(dX)^2 + b(dY)^2 + c(dX)(dY)$$

where a, b and c are numbers that vary from place to place on the grid. We changed the notation and wrote dz instead of Δz. The two quantities have the same meaning – the difference between two coordinates – but we are going to reserve using the d's for the special case in which these distances are small. This formula is true for any grid and a similar formula can be written down in more than two dimensions. For any curved surface, the set of numbers like a, b and c provide the rule for how to compute distances. Collectively, that set of numbers (like a, b and c) is called the 'metric' for the surface. Once we know the metric for our chosen coordinate grid, we can compute distances. A big part of general relativity is deriving the metric for a particular situation. This is what Schwarzschild did for the spacetime around a (non-rotating) star.

The Schwarzschild solution

We can now step back into spacetime and return to general relativity. Schwarzschild's solution tells us the metric in the vicinity of any spherically symmetric distribution of matter like a star or black hole. Using the spacetime grid that Schwarzschild used (more on that in a moment), the corresponding interval (distance) between two nearby events outside of a star or black hole is:

$$d\tau^2 = \left(1 - \frac{R_S}{R}\right) dt^2 - \frac{1}{\left(1 - \frac{R_S}{R}\right)} dR^2 - d\Omega^2$$

As in Chapter 1, the Schwarzschild radius is given by the formula:

$$R_S = \frac{2GM}{c^2}$$

where G is Newton's gravitational constant, M is the mass of the star and c is the speed of light. Pretty much everything that we want to know about non-spinning black holes in general relativity is contained within this one line of mathematics. We see that Schwarzschild chose a grid labelled with a time coordinate, t, and a distance coordinate R. Ignore the $d\Omega^2$ term for now, it won't be important and we'll explain why in a moment.

Because the spacetime is curved, it is not possible to find a single coordinate grid such that the flat (Minkowski) formula for the interval holds everywhere. That's the reason for the factors in front of dt^2 and dR^2. Conceptually, these factors are no different from those we had to introduce in the simpler two-dimensional case above: they encode the information about the curvature. The formula is telling us that the warping of spacetime at a particular location depends on how close that location is to the star and how massive the star is. The t and R coordinate grid chosen by Schwarzschild doesn't have to correspond directly to anything that can be measured using clocks or rulers. However, the R and t coordinates do have a physical interpretation, which will allow us to develop an intuitive picture of Schwarzschild's spacetime.

We've sketched what we might term a 'space diagram' of Schwarzschild's spacetime in Figure 4.4. The star sits at what we'll call the 'centre of attraction'. We've drawn two shells surrounding the star. Each shell is at a fixed Schwarzschild coordinate R from the centre of attraction (at $R = 0$). The R coordinate is defined in terms of the surface area of these shells. In flat space, the surface area of a sphere $A = 4\pi R^2$, and R is the distance to the centre of the sphere as measured by a ruler. In the distorted space around a star (or black hole) this is no longer true (for a black hole it's not even possible to lay a ruler down starting from the centre of attraction – the singularity). We can, however, always measure the area of spherical shells like those on the diagram and R is the radius a shell would have had if spacetime were flat. That's how Schwarzschild chose this coordinate.

No matter how the spacetime is distorted, the distortion must be the same at every point on these spherical shells. This is because Schwarzschild assumed perfect spherical symmetry when he derived his equation. Think of a perfect sphere; every point on the surface is the same as every other point. The $d\Omega^2$ piece in the equation deals with the distance between events on a particular shell. It is precisely the same as the piece we find in the metric used to calculate distances on the surface of the (spherical) Earth. We've discussed this in a bit more detail in Box 4.1. We would need this piece if we wanted to calculate the details of orbits around a star or black hole, but in what follows we'll always consider things moving only inwards or outwards. This will simplify matters while still capturing the important physics.

The Schwarzschild time coordinate also has a simple definition: t corresponds to the time as measured on a clock far away from the centre of attraction, where spacetime is almost flat. As we move inwards towards the star, spacetime becomes more curved

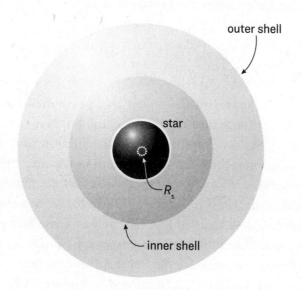

Figure 4.4. Schwarzschild 'space diagram'. The star sits at the centre. Two imaginary spherical shells surround the star.

and that's why we need the factors in front of dR^2 and dt^2. These factors are more important as we move inwards and are close to 1 far away. This makes sense because it means that far away from the star the interval is the same as it is in flat space.

The fact that Schwarzschild's coordinates have a simple interpretation far from the star allows us to understand what the curvature of spacetime means for the passage of time and the measurement of lengths closer in. Let's imagine we have access to a small laboratory that we can position anywhere in spacetime. Our laboratory has no rockets attached and is therefore in freefall. Inside the laboratory there is a watch to measure the passage of time and a ruler to measure distance. Our lab is small in both space and time, which means we can assume spacetime is flat inside the lab. Let's now locate the laboratory in the vicinity of the outer shell of Figure 4.4 and observe the watch ticking. If the ticks are so short that our laboratory stays at roughly the same R coordinate for the duration of the tick, Schwarzschild's equation tells us that the spacetime interval between ticks is:

$$d\tau^2 \approx \left(1 - \frac{R_S}{R}\right) dt^2$$

We've used an 'approximately equals' sign to emphasise that we're making an approximation. In this case, d$R \approx 0$ between the ticks so we can ignore the dR piece of Schwarzschild's equation.

This equation is the origin of our claim in Chapter 1 that time passes more slowly for an astronaut when they are close to a star or black hole. dt^2 is the time interval (squared) we would measure between the ticks of our laboratory watch as measured by a clock at rest far away from the star, where spacetime is not curved. The $(1 - R_S/R)$ factor corrects for the fact that this does not correspond to time in our laboratory, as measured by the laboratory watch. The curvature has distorted time and so we need more ticks of the distant clock for one tick of the laboratory watch. Time has slowed down at the location of the laboratory relative to far away. On the inner shell in Figure 4.4, R is smaller still. If we place our laboratory there, the number in front of dt^2 will be even smaller, and therefore watches on the inner shell will run even slower.

What about space warping? Imagine sitting in the laboratory at the outer shell and measuring the distance to a nearby lower shell using a ruler. If the lower shell is close by, the measured distance on the ruler between the two nearby shells is given by the second term in Schwarzschild's equation:

$$d\tau^2 \approx -\frac{dR^2}{\left(1 - \dfrac{R_S}{R}\right)}$$

Here dR would be the distance between the shells if space were flat. Since the factor $(1 - R_S/R)$ is now in the denominator, the distance between the nearby shells as measured by an observer on one of the shells is larger than it would have been in flat space. This means that space is being stretched and time is being slowed down as we get closer to a star.

To get a feel for the size of these effects, we can put the numbers in for the case of the Sun. The Sun's Schwarzschild radius is approximately 3 kilometres and its radius is approximately 700,000 kilometres. This gives a distorting factor of 1.000002 at the surface of the Sun. This means that, for two Sun-sized shells whose radii differ by 1 kilometre in flat space, the measured distance between them would be 2 millimetres longer than 1 kilometre. Likewise, an observer far away from the Sun would see a watch at the Sun's surface run slow by 2 microseconds every second, which is around a minute per year.

The Schwarzschild black hole: just remove the star

Schwarzschild's solution was originally used to study the region outside of a star or planet (the region inside the star is filled with matter and his solution is not valid there). The remarkable thing is that the same solution can also be used to describe a black hole. All we need to do is to ignore the star. Schwarzschild's solution then describes an infinite, eternal Universe in which the spacetime becomes more and more distorted as we head inwards towards the singularity at $R = 0$: a perfect eternal black hole.

We've sketched Schwarzschild's space without a star in Figure 4.5. The two imaginary shells we considered before are still there, but the star has disappeared, leaving only empty Schwarzschild spacetime. We've also drawn a shell at the Schwarzschild radius, which previously lay inside the star. Looking back at our equations, something very strange happens on the shell at the

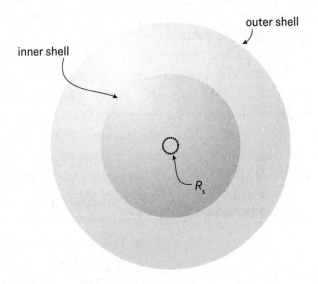

Figure 4.5. Schwarzschild 'space diagram' with the star removed. There is no matter anywhere.

Schwarzschild radius: the $(1 - R_S/R)$ factors are equal to zero. Even more dramatically, as we continue further inwards, these factors become negative. What does this mean? From the perspective of someone in freefall across the shell at the Schwarzschild radius the Equivalence Principle informs us that nothing untoward happens. And yet, from a distant perspective, the shell is a place where clocks stop and space has an infinite stretch.

To understand what is happening, it helps to draw some pictures. Before plunging in with Penrose diagrams, we can learn something from a spacetime diagram. As with the diagrams of flat spacetime we've already met, there are many ways to construct these diagrams (corresponding to different choices of grid). We will use Schwarzschild's coordinate grid since we have just seen that something interesting happens at the Schwarzschild radius. Figure 4.6 shows the light cones at each point in Schwarzschild spacetime. This should be contrasted with the corresponding diagram in flat spacetime. If spacetime is flat, the light cones are all aligned and point vertically upwards, but that is not the case in Schwarzschild spacetime. Far from the Schwarzschild radius, the light cones do look like those in flat spacetime, but as we approach the Schwarzschild radius the cones get narrower and narrower. At the Schwarzschild radius the light cones are infinitely narrow, which means an outgoing beam of light can only travel in the time direction and can never climb away from the hole.* Now we can appreciate that the Schwarzschild radius is also the event horizon: an outgoing beam of light emitted at the Schwarzschild radius stands still.

Inside the horizon, the light cones have flipped round. This is because the $(1 - R_S/R)$ factor has become negative, which means

---

* You might be inclined to think ingoing beams are also stuck forever on the horizon. This is the case from the perspective of someone far from the hole (for whom Schwarzschild time is their clock time). But it does not mean things cannot fall into the black hole from their own perspective. We shall have more to say on this in Chapter 5.

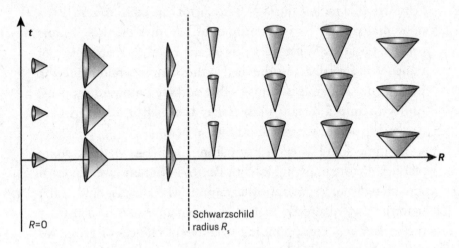

Figure 4.6. Schwarzschild spacetime. The *t* and *R* coordinates are those
used by Schwarzschild. Notice how the light cones tip over at values of
R smaller than the Schwarzschild radius.

the factor in front of dt^2 gets a minus sign and dR^2 gets a plus sign.
It is as if space and time have switched roles, but in fact what has
switched roles is our interpretation of the Schwarzschild *t* and *R*
coordinates. Because light cones open out around the
Schwarzschild *R* direction, this is the direction of 'time' for
anything inside the horizon, and the Schwarzschild *t* direction is
now 'space'. Since Schwarzschild coordinates correspond to measurements made using clocks and rulers for someone far from the
black hole, this means that what is time for someone inside the
black hole is space for someone far away, and vice versa. As we've
been at pains to emphasise, the coordinates we use don't have to
correspond to anyone's idea of space and time: to quote Einstein
again 'they do not have to have an immediate metrical meaning'.
The Schwarzschild *R* and *t* coordinates do happen to have a nice
interpretation far away from a black hole, but inside the horizon
their roles flip. The startling consequence is that an object inside
the horizon moves inexorably towards the centre of attraction at
R = 0, just as surely as you move inexorably towards tomorrow.

We haven't said much about the centre of attraction yet. Inside a black hole, this is the singularity, the 'place' where Einstein's theory and Schwarzschild's solution break down. The quotation marks are appropriate because the singularity isn't really a place in space. It is a moment in time: the end of time that lies in the future for all who dare to cross the horizon. Figure 4.6 illustrates very nicely that the singularity lies in everything's future inside the horizon, because all the light cones point towards it. It's also evident from Figure 4.6 that the singularity is not a point in space, which is what we are tempted to think when we look at Figure 4.5. The time and space role reversal means that it is an infinite surface at a moment in time. Let's explore this remarkable claim in more detail by plunging in and drawing the Penrose diagram for a Schwarzschild black hole.

The Penrose diagram for the eternal Schwarzschild black hole

The Penrose diagram for the eternal Schwarzschild black hole is shown in Figure 4.7. It is built from two portions: the diamond shape to the right corresponds to the universe outside the black hole. The triangle at the top corresponds to the interior of the black hole and the dividing line between the two is the event horizon. It is a 45-degree line because the horizon is lightlike, which means that light can 'get stuck' on it. The singularity is the horizontal line at the top edge of the triangle. It is horizontal because it corresponds to the inexorable future of anything that falls inside the horizon. To see all of this, recall that light cones are always oriented vertically upwards on a Penrose diagram, and worldlines always head into future light cones.

We have drawn a grid on the diagram, just as we did for flat spacetime. This grid is Schwarzschild's grid. In the diamond region, the roughly horizontal lines are lines of constant t and the roughly vertical lines are lines of constant R. The event horizon

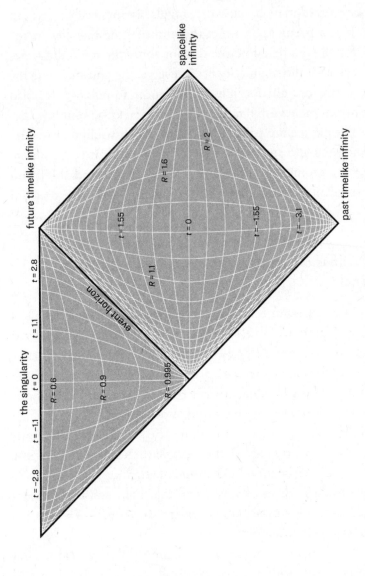

Figure 4.7. The Penrose diagram corresponding to the eternal Schwarzschild black hole. The grid corresponds to lines of fixed Schwarzschild coordinates (any similarity to the grid drawn on the Penrose diagrams in the last chapter is accidental).

lies at $R = 1$.* In the interior of the black hole, we can see the role reversal of space and time because the lines of constant Schwarzschild t now run vertically and the lines of constant Schwarzschild R are horizontal. Unlike Figure 4.6, the Penrose diagram is constructed such that the future light cones always point vertically upwards, which means that time is always up and space is always horizontal at every point.

A nice feature of this diagram is that we can use it to describe a black hole of any mass we want. The supermassive black hole in M87, for example, has a Schwarzschild radius of around 19 billion kilometres, corresponding to a mass of 6 billion Suns. An object at $R = 2$ would then be hovering 19 billion kilometres above the event horizon. If instead we want to describe a black hole with the mass of our Sun, an object at $R = 2$ would be hovering a mere 3 kilometres above the event horizon. The same thing works for Schwarzschild time too, with one unit of t corresponding to around 18 hours for M87* or (dividing by 6 billion) 10 microseconds for a black hole of one solar mass.

To get more of a feel for the Schwarzschild spacetime, we can classify the edges of the Penrose diagram, just as we did for flat spacetime. The two 45-degree edges on the right of the diamond correspond to past and future lightlike infinity. Only things travelling at the speed of light can come from or reach these places. The right-hand apex of the diamond, where these two edges meet, corresponds to spacelike infinity. The bottom and top apexes of the diamond are past and future timelike infinity. This is very similar to the Penrose diagram of flat spacetime. The new feature is the horizontal line at the top of the diagram, labelled 'the singularity'. We can gain a good deal of insight by enlisting the services of two more intrepid astronauts who are exploring the black hole

---

* This is because we've chosen to label R in units of the Schwarzschild radius. With this choice, an object accelerating to hold at a fixed distance of twice the Schwarzschild radius from the black hole will be represented by a worldline that runs along the $R = 2$ grid line.

in the centre of M87. Their worldlines are illustrated in Figure 4.8, which shows that both the astronauts begin their voyage of exploration at $R = 1.1$; it's as if we've plonked them gently into the spacetime. Blue is a very relaxed astronaut and chooses to do nothing at all. He has rocket engines, but he doesn't bother to switch them on and freefalls across the horizon and into the black hole. Red is more sensible. She immediately flicks the switch on her rocket engines and accelerates away from the black hole. Her acceleration is sufficient to escape the black hole's gravitational pull, and she later flicks the switch on her engine and coasts happily away to future timelike infinity.

Just as in Chapter 3, we have marked the astronauts' journeys with dots, and the spacing of the dots corresponds to one hour as measured on their watches. There are an infinite number of dots along Red's worldline, because she is immortal and lives into the infinite future. Things are very different for Blue, however. Undergoing no acceleration, he initially feels as if he is floating. In accord with the Equivalence Principle, there is no experiment he can do inside his spaceship to tell him he is in the vicinity of a black hole, but there is a shock in store in his future. After crossing the horizon, there are only 20 dots on Blue's worldline. At some point during the twentieth hour, something bad happens. The relaxed immortal's worldline ends. He meets the singularity. As you can see from the Penrose diagram, the singularity is unavoidable.

All the immortals in flat spacetime live forever, no matter how they move. In Schwarzschild spacetime, every worldline that enters the upper triangle must end on the singularity. Nobody is immortal once they travel beyond the horizon of a black hole. The interior of a black hole is a fascinating place; a Dantean wonderland where all hope would appear to be abandoned as space and time flip roles and the end of time awaits. But there is much more to say, so let's cross the horizon and explore.

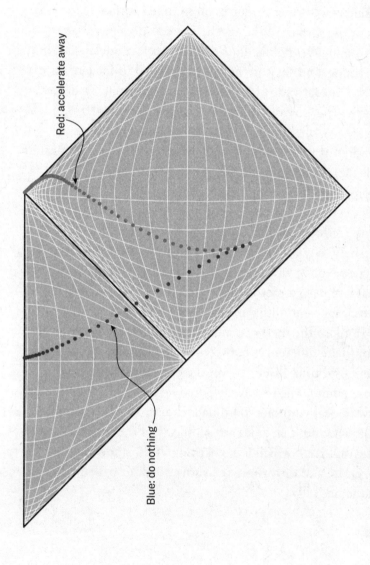

Figure 4.8. The journeys of Blue and Red in the vicinity of a Schwarzschild black hole. The dots lie on their worldlines and are 1 hour apart in the case that the black hole is the one in the centre of M87. Also on plate 5.

Red: accelerate away

Blue: do nothing

BOX 4.1. The surface of the Earth

A good way to see the difference between coordinate distances and ruler distances on a curved surface is to think about the surface of the Earth. The coordinates we often choose to label points on the Earth's surface are latitude and longitude, and they are not simply related to the lengths of rulers. The latitude of London is approximately 51 degrees North. The City of Calgary in Canada also sits at around 51 degrees North, and its longitude is 114 degrees west of London. If we ask a pilot to fly between the cities at constant latitude, the distance the aircraft travels will be approximately 5,000 miles, corresponding to a journey of 114 degrees in our chosen coordinate system. If we made the same 114-degree journey from Longyearbyen, the most northerly city on Earth sitting inside the Arctic Circle at 78 degrees North, we'd only travel around 1,600 miles. We therefore need a metric that translates coordinate differences into distances at different places on the surface.

Think of the metric as a little machine that takes coordinate differences between two points and spits out the real-world ruler distance over the curved surface between those points. The metric encodes two things: it understands how to deal with our coordinate choice, which is completely arbitrary, and it encodes the geometry of the surface – in this case a sphere – which is a real thing. This is how the curvature and distortion of a surface is dealt with mathematically.

5

INTO THE BLACK HOLE

In the film *Interstellar*, Matthew McConaughey dives into a black hole called Gargantua and emerges inside a multi-dimensional reconstruction of his daughter's bookshelves. That's not what happens in Nature.* But what is the fate of an astronaut who decides to embark on a voyage beyond the horizon into the interior of a black hole? We are now equipped to answer that question for black holes that do not spin, according to general relativity. In Chapter 6, we'll add some spin and explore the interior of what are known as Kerr black holes. This will allow us to embark on even more fantastical voyages into a wonderland of wormholes and other universes. But first things first.

For our purposes, we are going to recruit three more astronauts to join Red and Blue from the previous chapter in their exploration of the supermassive black hole in M87. Their journeys over spacetime are shown in Figure 5.1. We have marked out their positions as time advances using coloured dots.

---

* Our PhD student Ross Jenkinson has a different take on it: 'My interpretation was that the 5D beings picked him up in a 5D box and saved him from the black hole, carrying him through an unseen dimension, which they represented as him being able to travel through time as if it were a dimension in space. Analogous to picking up a flatlander in some tupperware as they fall through a 3D black hole.' That's also not what happens in Nature.

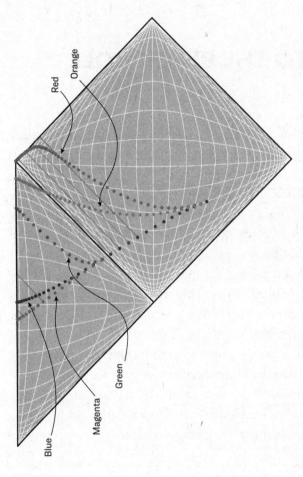

Figure 5.1. The Penrose diagram corresponding to the eternal Schwarzschild black hole. Five astronauts start out at rest at R = 1.1. Four of them head towards the black hole. Blue freefalls, Green freefalls with Blue but switches her rockets on after crossing the horizon and attempts to accelerate away from the singularity. Magenta also travels with Blue and Green until she reaches the horizon, after which she accelerates towards the singularity. Red accelerates away from the black hole from the moment her journey begins and succeeds in escaping to infinity. Orange also accelerates away from the black hole, but not enough to escape. Also on plate 6.

We followed Blue in the previous chapter. He sets out from rest at $R = 1.1$, freefalls into the black hole and ends up at the singularity. His worldline is marked in units of one hour by his watch and nothing unusual happens from his perspective at the event horizon – he sails through it oblivious. There are 20 dots on his worldline inside the horizon, which corresponds to almost a day inside the black hole before the end of time.*

Green begins her journey alongside Blue and freefalls towards the horizon with him, but on crossing the horizon she panics, shouts 'Burma'† and turns on her rocket engine in a vain attempt to escape. There are only 16 dots along her worldline once she crosses the horizon, which means that accelerating away has resulted in the end of time arriving sooner.‡

Magenta takes what might seem, at first sight, to be a more fatalistic view. She decides to fall together with Green and Blue until the horizon, and then she gently accelerates towards the end of time at 480g (which is five times less acceleration than Green). She presses play on Joy Division's 'Unknown Pleasures' and flicks the switch. Perhaps irritatingly for her, this lengthens her stay inside the horizon and she lives longer than Green – her worldline has 17 dots. The spacetime geometry inside the black hole is certainly counter-intuitive.

The maximum time anyone can spend inside the horizon of a black hole corresponds to someone who starts out from rest on

---

* If the black hole were one solar mass, Blue would have only 14 microseconds after crossing the horizon before reaching the singularity. If you want to explore the interior of a black hole, you should choose a big one otherwise the adventure will be over very quickly.

† This is an obscure Monty Python reference. Every popular science book should contain one.

‡ The acceleration depicted here for Green is a bone crushing 2,400g. If this were a solar mass black hole, Green would be experiencing 15 trillion g, which would be even more uncomfortable. It is just as well that our astronauts are immortal. To experience 1g, Green would have to dive into a black hole 2,400 times the mass of M87*. The most massive black hole known at the time of writing is ten times the mass of M87*.

the horizon and does absolutely nothing but fall freely to the singularity. Apathy pays. This corresponds to just over a day (28 dots) inside the horizon of the supermassive black hole in M87.

On passing through the horizon, Green and Magenta accelerate away from the apathetic Blue, who's probably relaxing to Miles Davis. Green accelerates away from the singularity and Magenta accelerates towards it. Blue sees Green recede into the distance above him in the direction of the horizon, and Magenta heads away in the direction of the singularity. They both get smaller and smaller from Blue's point of view as they disappear into the distance in opposite directions. So far so normal. It certainly looks like Magenta is heading towards her doom at the singularity and Green is doing her best to stay close to the horizon. None of this seems different to the way things would be in any other region of spacetime.

In Figure 5.1, we've drawn two light beams that Magenta shines out. The first is emitted just under nine hours after she passed through the horizon and the other at just under 14 hours. Drawing light beams like this is what we should do if we want to investigate what each of our astronauts actually sees. Remember that the beauty of Penrose diagrams is that the light cones all point vertically upwards and open out at 45 degrees. Notice that the earlier light beam intersects Blue's trajectory. That means Blue sees Magenta emit the beam of light (he receives the light at the point on his trajectory where the beam intersects it). Now for the fun bit. Look at the second beam Magenta shines out. It never intersects Blue's trajectory, and therefore Blue never sees Magenta turn on her torch to emit that second beam. In other words, Blue never sees Magenta's final moments, even though she is accelerating away from him. The light from Magenta's final four dots simply doesn't have time to enter Blue's eyes before he reaches the end of time. He would see her way down below him as the last light reached him before he winked out of existence.

If Blue turns around, he'll see Green way above him attempting to accelerate away from the singularity. Again, he'll reach the end of time before he sees Green end her days. What's interesting is that every astronaut has the same experience. Nobody ever sees anyone else hit the singularity. The reason is the horizontal nature of the singularity on the Penrose diagram. It is a moment in time, and we can never see events that are simultaneous with a moment in time. We always see things slightly in the past, because it takes light time to travel to our eyes. This means that nobody falling into the black hole sees anyone else reach the singularity before they themselves reach it – they quite literally never see it coming. If you are struggling to see this, imagine drawing 45-degree light beams all over the diagram. They'll tell you what each astronaut can and can't see.

The singularity doesn't arrive entirely unannounced to the unsuspecting astronauts though because as they approach the singularity they get stretched out by tidal forces: they get spaghet-tified. Standing on the Earth, the pull of gravity is slightly greater at your feet than at your head, but not so much that you notice that you are being stretched. The gravitational pull of the Moon on the Earth has a similar stretching effect, causing the more noticeable twice daily tides. We can see how tidal forces arise using the Penrose diagram in Figure 5.2.

The dotted lines correspond to the worldlines of two balls fall-ing into the hole. One ball starts out at $R = 2$ and the other a little closer in at $R = 1.8$. Once again, the dots correspond to regular ticks of a clock (imagine a clock glued to each ball). The dots are close together to make it easier for us to see the tidal effects (though harder to count the dots). The 45-degree lines correspond to a light beam bouncing back and forth between the balls. We can use the bouncing light as a ruler to measure the distance between the balls, just as you might use a laser tape measure at home if you enjoy putting up shelves. The numbers are the number of ticks between successive bounces as measured on the

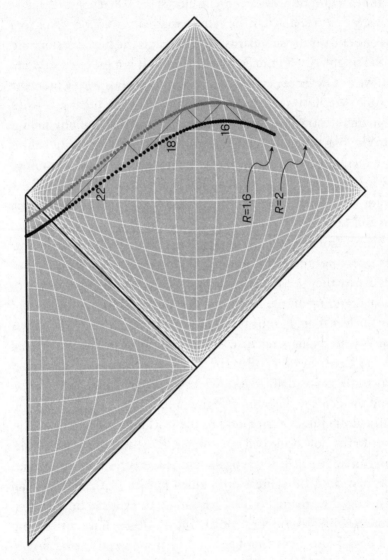

Figure 5.2. Spaghettification. Also on plate 7.

watch attached to the lower ball. These numbers correspond to the roundtrip travel time of the light pulses. The key thing to notice is that the time between bounces increases as the two balls fall towards the horizon. This means that the balls are moving apart as they fall towards the black hole.

Imagine now that you are falling into the black hole feet first. Your head and feet will try to move apart but, since they are connected by your body, you'll instead feel like you are being stretched. For the black hole in M87, the tidal effects at the horizon would be unnoticeable, but you'd begin to feel uncomfortable inside at around $R = 5$ million kilometres. At some point around 3 million kilometres your head would come off. You'd have been spaghettified. Closer to the singularity, your constituent atoms would be ripped apart. Even more dramatically, for a typical stellar mass black hole you'd be spaghettified before you even reached the horizon.

Let's now return to Figure 5.1 and ask what things look like to an observer who remains outside the black hole. Red sets out alongside Blue, Green and Magenta but wisely decides to switch on her engines in good time and accelerate away from the black hole. In good time is a 'relative' term here – she pulls 864g until Schwarzschild $t = 1.5$, at which point she decides enough is enough and switches off her rocket engine. Red's 864g acceleration should probably be accompanied by Kenny Loggins's 'Danger Zone'. You can see the moment when Red switches off her rockets because that's when her worldline takes an abrupt left turn and heads for the apex of the diamond. Red has escaped the black hole and, being immortal, her worldline will continue to future timelike infinity at the top apex of the Penrose diagram. Remember that nothing peculiar happens to anyone, at any point before they start to experience tidal forces and Red manages to avoid those. She does feel the acceleration of her rocket for the first portion of her journey, but once that's done, she floats around happily inside her spaceship as she coasts into infinity.

What does Red see as she observes her colleagues diving into the black hole? Before Blue, Green and Magenta reach the horizon we have drawn some more 45-degree lines. We can use these to confirm that Red sees the others move in slow motion and that this gets increasingly pronounced as they approach the horizon. To trace the journey of the light as it travels from Blue to Red's eyes, follow the 45-degree lines from Blue to Red's worldline. Can you see that two dots on Blue's in-falling worldline correspond to very many dots on Red's worldline? This means that Red experiences many hours for every hour that Blue experiences. Dramatically, Red never even sees anybody pass through the horizon because light emitted very close to the horizon heads off at 45 degrees and only reaches her in the far future. This means she continues to receive light from the other astronauts forever. She sees in-falling objects move ever slower as they approach the horizon, until they eventually freeze there. In principle, she can see everything that ever fell into the black hole.

There is a second important consequence of the fact that Red sees the in-fallers in ever-increasing slow motion as they approach the horizon. As we've emphasised, the slowing down of the astronauts and watches is not specific to astronauts and watches. Everything slows down, from the rate of ageing of the cells in the astronauts' bodies to the inner workings of atoms. Time is distorted, and that means every physical process is distorted too. This includes light. Light is a wave and has a frequency, just like sound or waves on the surface of water. Water waves are perhaps the best way to picture a wave if you aren't familiar with the terminology. If you throw a stone into a still pond, ripples radiate out from the stone. Standing still in the pond, you'll feel a series of peaks and troughs as the wave passes by. The distance between two peaks is known as the wavelength, and the number of peaks that pass by per second is known as the frequency. For visible light, we perceive the frequency as colour. High-frequency visible light is blue and low-frequency visible light is red. Beyond the visible at the low-frequency end of the spectrum are infrared light, microwaves and radio waves. Beyond the high end lie ultraviolet light, X-rays and gamma rays.

A distant observer such as Red sees the in-fallers by the light they emit, and the frequency of that light reduces as time slows down. The images of the in-falling astronauts therefore become redder as they approach the horizon, and ultimately fade away as the frequency drops out of the visible range and into the micro-wave and radio bands beyond. This effect is known as redshift. Red sees her in-falling colleagues freeze *and* fade away as they approach the horizon.

We will end our investigation of the Schwarzschild black hole by commenting on an apparent paradox that's confused many people in the past. We now understand that nobody outside the black hole ever sees anyone fall through the horizon. But we have also said that astronauts do fall through and can see each other inside the horizon. How does that work? Won't it be the case that astronauts falling towards the horizon should never see their colleagues ahead of them fall through? Worse still, if they go in feet first, won't their feet appear to freeze below them? Will they fall through their own feet? The answer is that nobody sees anyone else fall through the horizon until they themselves are inside. Even more bizarrely, nobody even sees their feet cross the horizon until their eyes have crossed it. Let's look at the Penrose diagram to work out how this can be the case and why, in fact, it is not in the least bit bizarre.

In Figure 5.1 there is an astronaut we haven't met before who we've called Orange. She is listening to Monty Python. She starts out with our other astronauts and attempts to avoid falling in but, being a little silly, she accelerates away from the black hole too slowly and ultimately crosses the horizon. As she approaches the horizon, she sees Blue, Green and Magenta approaching the horizon in slow motion. But, because the horizon is also a 45-degree line, she doesn't see anyone cross the horizon until the moment

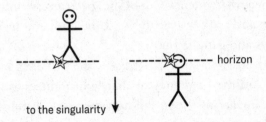

Figure 5.3. Orange not falling through her own feet.

her eyes cross it. This also applies to her own feet. She doesn't see them cross the horizon until the moment her eyes cross it. This sounds weird, as if everything is piling up on the horizon and that Orange falls through some kind of ghostly mirage of everything that ever fell into the hole.

But there is nothing unusual here. Orange no more falls through her own feet than you fall through your own face when you walk towards a mirror. Let us explore that sentence in more detail.

Figure 5.3 shows Orange at two moments in her journey across the horizon; the moment when her feet cross and the moment when her eyes cross. The flash indicates light emitted from her foot at the moment it reaches the horizon. This light remains stuck on the horizon while she falls. From Orange's perspective, the horizon and the light whizz past her eyes. She sees her feet, but only when her eyes reach the horizon. But this is what always happens when you look down at your feet: the light from them travels up to your head and you see them after the light has been emitted.

What about the light from everyone else who crossed the horizon? All that light is simply waiting around on the horizon until Orange's eyes pass by and collect it. Again, there is nothing unusual about that. Presumably you aren't puzzled by the fact that you can see distant cows and nearby cows standing in a field at the same time.

To make these unfamiliar ideas clearer we'll introduce another way to think about spacetime around a black hole: the river model. The river model was so-called by Andrew Hamilton and Jason Lisle and it has an impeccable pedigree.[18] It was formulated by Allvar Gullstrand in 1921, who had previously won the 1911 Nobel Prize in Physiology or Medicine for his work on the optics of the eye. French mathematician Paul Painlevé discovered the model independently in 1922, in between his two stints as Prime Minister of France. (Given the intellectual abilities of some recent holders of high office in the United Kingdom and elsewhere, the preceding sentence assumes an almost comedic quality.) In 1933, Georges Lemaître showed that the river model correctly describes a Schwarzschild black hole but with a different choice of grid.

In the river model, we are entitled to think of a Schwarzschild black hole by analogy with water flowing into a sink hole, as illustrated in Figure 5.4. The water represents space, which flows into the hole at ever-increasing speed. Light, and indeed everything else, moves over the flowing river of space in accord with the laws of special relativity. We might imagine our astronauts swimming around in the flowing river of space. Far from the black hole, the flow is sedate and they can easily swim away upstream. As they approach the black hole, the flow gets faster and they find it increasingly difficult to escape. At the horizon, the flow reaches

Figure 5.4. The river model of a black hole. Also on plate 8.

the maximum speed that anything can swim (the speed of light), and since nothing can travel faster than light, this is the point of no return. Inside the horizon, the river flows faster than the speed of light and gets ever faster as it approaches the singularity. Anything that strays beyond the horizon will be caught up in the superluminal flow and inexorably swept to its doom. On the horizon, something swimming radially outwards at the speed of light will go precisely nowhere. It will remain frozen forever on the horizon.

This picture makes Orange's experience on crossing the horizon very clear. She is stationary in the river of space, but the river is flowing across the horizon, sweeping her inwards. Particles of light (photons) emitted from her feet will head outwards at the speed of light as normal, but because the river is flowing inwards at the speed of light, the photons from her feet that will encounter her eyes remain frozen at the horizon. Orange's head is swept across the horizon in the river at the speed of light, where she meets these photons and sees her feet. Thus, the photons from her feet enter her eyes after they've been emitted, travelling at precisely the speed of light. If you look down at your feet now, this is also exactly what happens. As she crosses the horizon, Orange experiences the world precisely as you are doing now.

We can invoke the river model to visualise other phenomena we have previously described using the Penrose diagram. Tidal effects arise because the river flows faster closer to the hole, which will cause two astronauts swimming at the same speed but at different radial distances to drift apart as they fall inwards. On our Penrose diagram, we charted only a single spatial dimension – the radial direction. The river model is two-dimensional, and that allows us to see some additional effects. Because the flow is converging inwards to the singularity, objects close to the hole will be squeezed in the tangential direction as well as being stretched in the radial direction. Two astronauts will get closer together tangentially as they head inwards but be pulled apart radially, experiencing a kind of double spaghettification. Heading towards a black hole, feet first, you get thinner and longer.

We can also picture why a distant observer far from the hole never sees anything cross the horizon. Think of photons as fish. Suppose someone in a canoe heading towards the horizon drops fish overboard once every second by their watch. The fish swim away upstream to a boat far away in comparatively still water. At first, the fish can swim upstream easily and arrive at the boat close to one second apart. But as the canoe drifts closer to the horizon, the fish struggle to swim away against the quickening flow and so the arrival time of the fish at the boat increases. This is the redshift effect we discussed above. On the horizon, the fish dropped from the canoe enter a river flowing at the maximum speed they can swim. They never escape upstream to reach the boat, and so the observer on the boat never sees the canoe cross the horizon.

In this chapter, we have explored the topsy-turvy world in and around a Schwarzschild black hole, and we have learnt what it feels like to jump into a black hole or to watch someone as they jump into one. Now, it is time to explore further, and to introduce a feature of general relativity beloved of science fiction writers and which may ultimately prove to be a key idea if we are to understand what space really is. Wormholes.

BOX 5.1. Small. Far away. Why does an astronaut see their colleagues as far away when they cross the horizon if the light from them is frozen on the horizon?

Let's imagine that Orange is falling through the horizon feet first. In the river model, we would picture her as floating in the river with her feet pointing downstream. At the moment her feet reach the horizon, a pair of light-speed fish set off from her feet. Let's call these light-speed fish 'photons' because that's what they represent. These particular photons

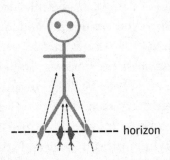

Figure 5.5. Photon-fish from Blue's feet enter Orange's eyes at a smaller angle than the photon-fish from Orange's own feet. Both enter the eyes at the same time, but Orange sees her own feet to be normal-sized and Blue to be distant and small. Also on plate 8.

are travelling at just the right angle to enter her eyes. There will be photons heading out in all directions from her feet of course, just as there are photons reflecting off your feet now in all directions, but only those that are heading in the correct direction will enter your eyes.

The photons we're considering are emitted at just the correct angle such that they will have moved inwards to meet Orange's eyes as those eyes pass by. It helps to think about the photons as fish swimming against the river, which is flowing vertically downwards in Figure 5.5. If they swim vertically at the same speed as the river, they will miss Orange's eyes. But if their path is tilted slightly inwards, they'll head inwards.* That is what they must do to reach Orange's eyes.

At the moment Orange's eyes reach the horizon they also encounter the photons coming from Blue's feet that were trapped on the horizon when he fell through. These photons have been there for longer than the photons from Orange's

---

* This means that the photons reaching Orange's eyes were actually emitted from her feet ever-so-slightly before they crossed the horizon.

feet and they've therefore had more time to make their way to meet Orange's eyes. This means that the photons from Blue's feet that happen to be at the right position to enter Orange's eyes as she passes by must have been emitted at a steeper angle, closer to the vertical, than the photons from her own feet. This means that Orange will see Blue to be smaller and, therefore, far away, because the size we perceive something to be is determined by the angular spread of the light arriving on our retina. For example, if we look out into a field, the distant cows look smaller than the ones nearby because they subtend a smaller angle.

Figure 5.6. Small. Far away.

6

WHITE HOLES AND WORMHOLES

Penrose diagrams bring infinity to a finite place on the page, and in Chapter 3 we explored how the different types of infinity are depicted at the edges and points of the diamond-shaped representation of flat spacetime. To refresh your memory, we've drawn the diagram for flat spacetime again, on the left of Figure 6.1. The upper and lower vertices of the diamond represent the distant past and the far future for anything or anyone who travels along timelike worldlines. We called these past and future time-like infinity. The worldlines of immortals begin and end there. Eternal light beams begin their journeys on one of the bottom edges and end on the opposite top edge. These are past and future lightlike infinity. All infinite 'now' slices of space stretch from the left-hand vertex to the right-hand vertex of the diamond. These are spacelike infinity. Every vertex and edge of the Penrose diagram represents infinity in one form or another.

Now look at the Penrose diagram for the eternal Schwarzschild black hole shown on the right of Figure 6.1. Just like the diagram for flat spacetime, this represents an infinite spacetime. We might expect that every vertex and edge of the diagram should lie either at infinity or the singularity. But this is not the case. What does the entire left-hand edge of the Penrose diagram represent? It doesn't lie at infinity but at an R coordinate of 1. This is the horizon of the black hole. Up to now, we've been focusing on the part of the

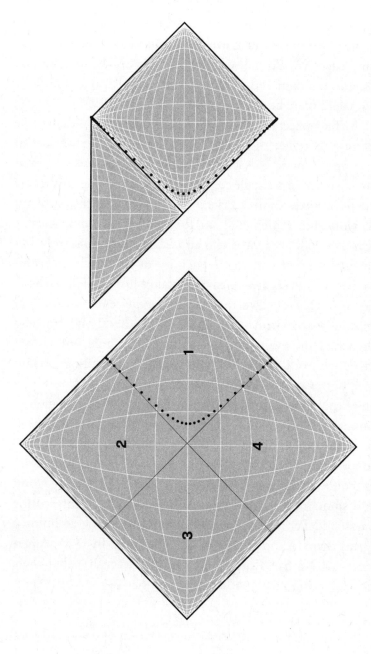

Figure 6.1. The Penrose diagrams for Minkowski spacetime (left) and the eternal Schwarzschild black hole (right). Rindler is accelerating through the spacetime on the left and Dot is hovering outside the black hole (at $R = 1.01$) on the right.

horizon that runs along the top left edge of the diamond-shaped region, beyond which lies the interior of the black hole, because this is the gateway for our brave astronauts. But what about that left-hand edge? We did not worry about it before because nothing travelling in the diamond can cross it, but since it does not lie at infinity, could there be something lurking beyond?

On the flat spacetime diagram of Figure 6.1, we've drawn the worldline of Rindler, the ever-accelerating astronaut we met in Chapter 3. He found himself hemmed in by horizons and lived out his existence in a smaller region of spacetime than his fellow immortals by virtue of his acceleration. Now look at the astronaut whose worldline is depicted on the Schwarzschild spacetime diagram. Let's call her Dot. She is also accelerating constantly, but in the curved spacetime in the vicinity of the black hole, this means that she hovers at a constant distance just outside the horizon at $R = 1.01$. Nevertheless, her experience inside her accelerating spacecraft is very similar to Rindler's. She can send signals across the event horizon of the black hole but cannot receive signals from inside. Likewise, Rindler can send signals into region 2 but cannot receive signals from it. For Rindler, we also recognise the presence of a horizon isolating him from region 4. He cannot travel to region 4, but he can receive signals from there. What is the meaning of the lower left-hand edge of Dot's diamond? Is the same true for her? Can she receive signals from across the lower horizon, and if so where are they coming from? There is something strange about this line. It is the edge of the Schwarzschild Penrose diagram, but it does not lie at infinity. Why can't there be something on the other side? In 1935, Albert Einstein and Nathan Rosen were the first to realise that there can be something on the other side.* In their words: 'The four-

---

* As early as 1916, in his paper, 'Contributions to Einstein's Theory of Gravitation', Ludwig Flamm had anticipated much of the Einstein–Rosen solution, but he does not appear to have identified the 'bridge'.

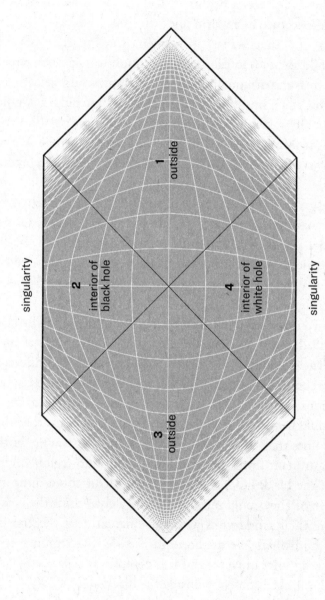

Figure 6.2. The Penrose diagram of the 'maximally extended' Schwarzschild spacetime. The grid lines correspond to so-called Kruskal–Szekeres coordinates as described in Box 6.1.

dimensional space is described mathematically by two ... sheets ... which are joined by a hyperplane ... We call such a connection between the two sheets a bridge.'[19] Today, this 'Einstein–Rosen Bridge' is also known as a wormhole.

It turns out that we've only been drawing half of the Schwarzschild solution to Einstein's equations. It is part of a larger space known as the maximally extended Schwarzschild spacetime, which is a bit of a mouthful. The eternal Schwarzschild black hole is to maximally extended Schwarzschild spacetime as Rindler's quadrant is to Minkowski space; a piece of a larger whole. We've drawn the maximally extended Schwarzschild spacetime in Figure 6.2.

The most striking things about Figure 6.2 are the entirely new regions of spacetime that have appeared: regions 3 and 4. Given that region 1 was the entire infinite universe outside of the black hole and region 2 was the region inside containing the end of time, you may be forgiven for wondering what regions 3 and 4 could possibly be. Let's explore.

Being a Penrose diagram, time runs upwards and all light rays travel at 45 degrees. We can therefore draw light cones at any point on the diagram and immediately see how the different regions are connected to each other. Things can travel from region 4 into regions 1, 2 and 3, but the reverse is not possible. Region 3 is inaccessible from region 1 and vice versa. This means that the 45-degree lines that cross in the middle of the diagram are horizons. An astronaut from region 1 could jump into region 2, the interior of the black hole, and another astronaut could jump in from region 3. They could meet up to have a chat inside the black hole before their rendezvous with the singularity at the end of time (the horizontal line at the top). We see that region 3 is a whole other infinite universe and it is completely separated from region 1, but linked somehow inside the black hole.

Another striking new feature is the horizontal line at the bottom of the diagram, which is also a singularity. Nothing ever falls into this singularity, and anything inside region 4 that lives long enough must eventually cross one of the horizons and enter regions 1 or 3. Anyone in 'universes' 1 or 3 could therefore encounter stuff that emerges across the horizons from region 4. This is the reverse of a black hole. It is called a white hole. The black hole lies in the future for astronauts in the two infinite universes, and they may or may not choose to fall into it. Conversely, the white hole lies in the past for these astronauts. They may receive signals from it, but they can never visit it. This is a rather dramatic turn of events.

We've referred to this diagram as the maximally extended Schwarzschild spacetime. The term 'maximal' has a technical meaning which, unlike many technical terms, is quite illuminating. The astronauts we've followed on their journeys around and into the black hole are immortal, which means that their worldlines should be infinitely long unless they hit a singularity. They live forever unless they cross the horizon of the black hole. This means that their worldlines must begin and end at infinity or on a singularity. A spacetime is maximal if it has that property. The spacetime representing the eternal Schwarzschild black hole in Figure 6.1 does not have that property because we can draw a worldline that enters the diagram on the left-hand edge. The maximally extended Schwarzschild spacetime is different. Every edge of the diagram lies either at infinity or on a singularity. For the Schwarzschild spacetime, this diagram is all there can be.

BOX 6.1. Kruskal–Szekeres coordinates

The grid we've drawn on Figure 6.2 is different to the Schwarzschild grid we've been using so far. Remember that we can choose any grid we like. Nature has no grid. These grid lines are marked out using Kruskal–Szekeres coordinates, discovered by Martin Kruskal and independently by George Szekeres in 1960.* The Kruskal–Szekeres grid lines correspond to spacelike (roughly horizontal) and timelike (roughly vertical) slices of the spacetime. Notice that the bunching up of the Schwarzschild grid lines at the horizons is avoided in Kruskal–Szekeres coordinates, which is more in line with the experience of time for astronauts who fall through the horizon without noticing anything strange. That said, it's worth remembering that the behaviour of Schwarzschild time at the horizon does tell us something important – that distant observers outside of the black hole see in-falling objects freeze on the horizon. To emphasise again, the coordinate grid we choose to locate events is a free choice, and different coordinate grids are more or less useful to different observers. The Schwarzschild grid is useful in describing the experience of observers outside the black hole because the Schwarzschild time coordinate corresponds directly to something measurable – it is the time as measured on clocks far away from the black hole. Kruskal–Szekeres time does not have that interpretation, but if we want to think about travelling across the horizons, the Kruskal–Szekeres grid is better.

Into the wormhole

Let's now explore the connection between the two universes in the maximally extended Schwarzschild spacetime. So far, we've mostly visualised spacetime using Penrose diagrams that represent a single dimension of space. These diagrams are a great way of visualising the relationship between events in different regions of spacetime

---

* Kruskal told John Wheeler about his coordinates that covered the entirety of the maximally extended Schwarzschild spacetime smoothly but didn't bother to publish the idea. Wheeler wrote up a short paper on the matter and sent it for publication with Kruskal as sole author, originally without Kruskal's knowledge. George Szekeres discovered the same coordinate system, also in 1960.

– who can influence what and when – but they are not so good for visualising the curvature of spacetime. We can construct a more intuitive visual picture by using what are known as 'embedding diagrams'.

A good way to understand embedding diagrams is to return to the surface of the Earth. We are going to erase from our minds the idea that the Earth is a sphere in three dimensions and think *only* of its surface, which is two-dimensional. Picture the surface as a kind of flatland, like the one we encountered in Chapter 2. Flat Albert and his flat companions are now busy doing all the things that geometers do to make themselves happy. They draw circles on the surface and calculate the value of π, which they will discover to be different to the value in Euclidean geometry. They will discover that if they travel far enough over this flatland, they'll arrive back where they started. If they have drawn a Mercator projection map, they will associate the points along the left and right edges with each other. They may also come to understand that they have introduced a great deal of distortion at the top and bottom of the map by their coordinate choice and will no doubt be motivated to find some new, complementary coordinates to better understand the polar regions. The important point is that all these observations and properties could be a feature of a two-dimensional universe with no third dimension. We are three-dimensional beings, and we can visualise a third dimension. As a consequence, we notice that there is an elegant way of representing this geometry by 'embedding' it in three dimensions. Flatland would then be represented as the curled-up surface of a sphere. The important thing is that the third dimension is not necessary and need not even exist. The three-dimensional space into which we imagine flatland to be embedded could be a hypothetical space (sometimes referred to as a hyperspace). Flat philosophers may enquire as to whether a third dimension really exists or not, but flat navigators will not care one way or the other. We 3D beings can use this third (hypothetical) dimension in our imaginations to visualise the curvature of the two-dimensional surface of flatland and gain a new perspective on the geometry. Before some flat-earther misconstrues this analogy, let us make it very clear that the Earth is actually a sphere in three-dimensional space. This is merely an analogy to help our understanding, and hopefully theirs. The point we wish to emphasise is that the 'curvature' observed by Flat Albert and his flat companions could be an intrinsic property of their two-dimensional space and does not require the existence of a third dimension. In our real Universe, the 'curvature' of our four-dimensional spacetime (which we experience as gravity) is not, as far as we can tell, a result of us living on a surface that is curved into a real fifth dimension. We don't think we are like the flatlanders, oblivious to some higher-dimensional universe into which our spacetime is curved.

In this sense, the word 'curvature' is a little misleading in general relativity because it encourages us to imagine a surface curving into an extra dimension. But curvature is a quantity that can be calculated directly from the metric with no reference to 'extra' dimensions at all. John Wheeler managed to say everything we've just said in the three-word title of a section in his book with Edwin F. Taylor:[20] 'Distances Determine Geometry'. The authors ask us to imagine a 'fantastically sculpted iceberg' floating on a 'heaving ocean'. To map its curving shape, we can imagine driving thousands of steel pitons into the ice and stretching strings between them. Then we note the positions of the pitons* and the lengths of the strings down in a book. This book contains all the information necessary to reconstruct the geometry of the iceberg, including the curvature of the surface. In spacetime, the pitons are the analogue of events – 'the steel surveying stakes of spacetime'. The distances between nearby events are the intervals. The book is the metric. Nowhere is there any reference to an extra dimension into which the iceberg is curved.

---

* We might arrange them in a rectangular grid.

Given that we are three-dimensional beings, we can use our imaginations to picture the curvature of two-dimensional spatial slices of spacetime, just as we could imagine the geometry of flat-land as the surface of a sphere. This is the beauty of embedding diagrams.

Before we head into the black hole, let's warm up by looking at spacetime in the vicinity of the Earth. Outside the planet, this will be described by the Schwarzschild metric which has three dimensions of space and one dimension of time. Imagine taking a slice of space through Earth's equator at a moment in time. In the language of relativity, this will be a two-dimensional space-like surface. On the left of Figure 6.3 we've drawn a Penrose diagram with such a slice through it. Earth sits at point O, and the slice runs from the Earth to X. If Earth wasn't there, this would be the Penrose diagram for flat spacetime. We've represented the slice OX through flat spacetime by a straight line at the top of the diagram. If we spin this line around O, we'll generate a sheet of two-dimensional space (the slice through the equator) centred on O. This is our embedding diagram, and for two-dimensional flat (Euclidean) space it looks like a sheet of graph paper.

If we now place the Earth at O, spacetime will be curved and the curvature outside the planet will be described by the

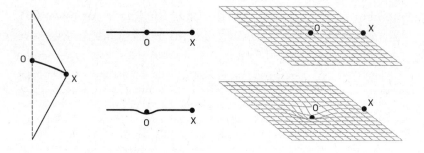

Figure 6.3. Representing a spacelike slice through spacetime.

Schwarzschild metric. To an astronaut in space close to the Earth, the curvature could be detected by making measurements of the distance between neighbouring events using a ruler, just as the surface of Wheeler's iceberg can be described using the lengths of pieces of string stretched between steel pitons. If you recall, the distance measured by the astronaut's ruler between two events, one slightly closer to Earth than the other, would be larger than expected had the space been flat. As for the case of 'curved' flatland, we could interpret this distortion with no reference at all to an imaginary extra dimension. Or, we could ask what shape the slice of space would have to be to produce the measured distortion *if it were curved into an extra dimension.* This is the grid we've sketched on Figure 6.3. From this perspective, the Earth makes a dimple in the fabric of space. Now let's construct some embedding diagrams to explore the geometry of the black hole.

In Figure 6.4 we've drawn five spacelike slices through maximally extended Schwarzschild spacetime. They all span the diagram from X to Y (the two spacelike infinities). These slices are snapshots of the geometry at different moments in time,* with earlier times towards the bottom of the diagram and later times towards the top. Let's focus first on the slice labelled (from right to left) YJIHX. We've drawn this slice as a line in Figure 6.5, just as we did for the spacetime around the Earth in Figure 6.3. The circles are encouraging you to think about the surface generated when we sweep the line around, but hold that thought for a moment and concentrate on the line. The line is flat towards Y because space is flat far away from the black hole. As we move

---

* These slices do not correspond to constant time slices using an array of identical clocks all at rest with respect to each other. It is a feature of flat spacetime that such a network of clocks can be conceived of, but the warping of spacetime makes it impossible to arrange in general. Rather they are slices of constant 'Kruskal' time. Nevertheless, the slices are spacelike in the sense that they have the property that no object can travel along any of the five curves (they are everywhere tilted at less than 45 degrees to the horizontal). These slices through spacetime are the best we can do to define the notion of a snapshot of space at some moment in time.

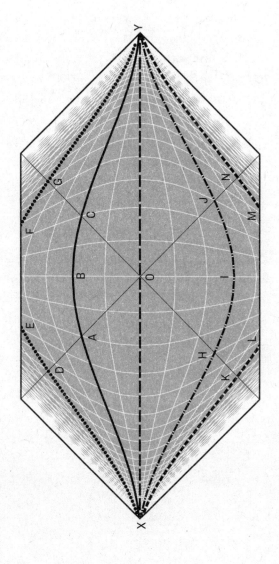

Figure 6.4. Five different slices through maximal Schwarzschild spacetime. Each slice can be regarded as all of space at a moment in time.

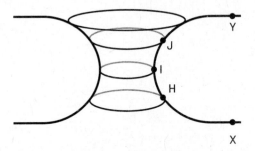

Figure 6.5. An embedding diagram of the spacelike slice YJIHX through the eternal Schwarzschild black hole, as described in the text. We can see the wormhole.

Figure 6.6. A wormhole in flatland.

inwards from Y to the event horizon at J, space starts to curve. So far so normal. On crossing the horizon, however, the line continues bending around until it crosses the second horizon at H. It then flattens out again as it approaches X. Now we can spin this line around as we did in Figure 6.3, and then we see what this interesting geometry corresponds to. Remarkably, we have two flat regions of space joined by what John Wheeler called the throat of a wormhole and what Einstein and Rosen called a bridge. The flat regions can be thought of as two separate universes linked by a wormhole.

A more artistic rendering of the wormhole is shown in Figure 6.6. Thinking in only two dimensions of space, we can imagine

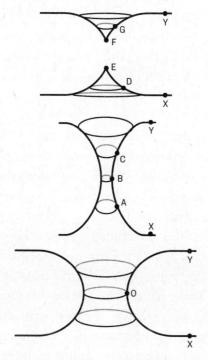

Figure 6.7. Embedding diagrams of three other spacelike slices from Figure 6.4. Time increases from bottom to top. We can see how the wormhole stretches open and snaps, leaving two disconnected universes. Nothing can travel through the wormhole before it snaps off.

Flat Albert and his friend sliding around the black hole. There is an infinite space inside. We humans can see how this works because we can picture the space curving in a third dimension, but for the flatlanders the idea would seem very strange. Similarly, you could imagine wrapping your hands around a low-mass maximally extended Schwarzschild black hole. Its horizon would be a tiny perfect sphere, but inside your cupped hands would reside an infinite other universe.

In Figure 6.7 we've drawn three more of the embedding diagrams representing the spacelike slices shown in Figure 6.4. We can see how the horizons move apart as the wormhole lengthens and eventually breaks to expose the singularity. The slice at the bottom occurs at earlier time than the slice at the top, which means that the wormhole evolves in time.* It is this evolution that prevents anything from travelling through the wormhole (see Box 6.2 for more detail). We don't need to draw wormholes to appreciate that nobody can travel from region 1 to region 3, and vice versa, which is what a journey through the wormhole would entail. That much is evident from the Penrose diagram since there is no line you can draw at an angle less than 45 degrees to the vertical that connects these two regions. However, the embedding diagrams of the wormhole provide a lovely picture of how the evolution of the wormhole renders travel through it impossible.

Figure 6.8 is a visualisation of the geometry of the maximally extended Schwarzschild spacetime together with an astronaut as he falls into the black hole. The wormholes are constant (Kruskal) time embedding diagrams. In the top left picture, the astronaut is approaching the horizon and the wormhole is open, connecting the two universes together. On the top right, the astronaut is about to pass through the horizon. He is still in region 1, but the wormhole has already passed its maximum diameter and is beginning to close. In the next image, he has crossed the black hole's

---

* We should really say 'Kruskal time' here, as described in Box 6.1.

Figure 6.8. An astronaut (the dot) falling into a Schwarzschild black hole. Time advances from the top left to bottom right. Notice how the wormhole grows and pinches off before the astronaut can reach the other side. Also on plate 9.

horizon and the wormhole is pinching closed, which it has done by the time of the second image on the bottom row. We see that the astronaut cannot traverse the wormhole because it pinches shut before he can pass through it. None of this is peculiar to any particular astronaut or how they manoeuvre on their journey into the black hole. The slamming of the door between universes has nothing to do with the details of the journey, and there is nothing anyone can do to change it. This story is written entirely within the Schwarzschild metric, the unique spherically symmetric solution of Einstein's equations. How wonderful.

BOX 6.2. Evolving wormholes

Let's start by writing down the Schwarzschild metric again:

$$d\tau^2 = \left(1 - \frac{R_S}{R}\right) dt^2 - \frac{1}{\left(1 - \frac{R_S}{R}\right)} dR^2$$

The $(1 - R_S/R)$ terms in front of the time and space coordinates tell us about the geometry of the spacetime – how it deviates from flat. These terms do not depend on time *outside the horizon*, which means the geometry does not change as t changes. The words *outside the horizon* are important. Inside the horizon, the space and time coordinates flip around such that the Schwarzschild R coordinate takes the role of time. If you recall, a key feature of life inside the horizon is that everything is compelled to move to smaller and smaller R, just as outside everything is compelled to move forwards in time. Why? Because the interval must always be positive along the worldline of anything with non-zero mass. This means that dt^2 cannot be zero outside the horizon and dR^2 cannot be zero inside the horizon. The ticking of time drives us forwards in t outside the horizon and forwards in R inside the horizon. This is why R is the time coordinate inside the horizon. But the $(1 - R_S/R)$ terms depend on R, which means that inside the horizon the geometry is compelled to change, just as inexorably as we are compelled to journey towards tomorrow. This is why the spacetime geometry is dynamic inside the horizon. It changes, and in such a way that not even light can make it through the wormhole.

Although travel from one universe to the other is impossible because the wormhole pinches shut, we have noted that it is possible for someone to jump into the black hole from region 1 and meet up with someone jumping in from region 3 before they both end up at the singularity. That is easy to see using the Penrose diagram. Moreover, someone inside the black hole in region 2 can see things in both regions 1 and 3 because they can receive signals from them. That is also evident from the Penrose diagram. This means that, in the moments between jumping into the black hole and hitting the singularity, our intrepid astronauts would be able to peer through the wormhole and see the universe on the other side.

We must of course ask whether any of this might play out in our Universe. Sadly, the answer appears to be 'probably not', at least for the case of astronauts attempting to travel between universes. That is because the maximally extended Schwarzschild spacetime does not correspond to the geometry of spacetime created by the gravitational collapse of a star. Rather, the Schwarzschild solution is only valid in the region of empty space outside of the star. The maximally extended Schwarzschild spacetime in Figure 6.2, replete with wormhole and black and white holes, would be the correct description of a non-spinning, eternal black hole. We are not aware that such things exist.

Why 'probably not'? Because wormhole geometries are valid solutions of Einstein's equations. In 1988, Michael Morris, Kip

Thorne and Ulvi Yurtsever explored the possibility of keeping the wormhole open.[21] 'We begin by asking whether the laws of physics permit an arbitrarily advanced civilisation to construct and maintain wormholes for interstellar travel?' These wormholes would not be constructed by collapsing stars, but could conceivably be 'pulled out of the quantum foam and enlarged to classical size and stabilised, potentially, by quantum fields with a negative energy density'. This is great fun but very speculative. As the authors state, such a wormhole would be a time machine, and the consequences of such things existing are disturbing. 'Can an advanced being measure Schrödinger's cat to be alive at an event P (thereby collapsing its wave function into a live state), then go backwards in time via the wormhole and kill the cat (collapse its wave function into a "dead" state) before it reaches P?' Putting the fun aside, these ideas have resurfaced in the search for a resolution to the black hole information paradox. In particular, the idea that microscopic wormholes could be part of the structure of spacetime is part of the ER = EPR hypothesis we'll meet in the final chapters. So maybe there are time machines in our Universe after all. In any event, the maximal Schwarzschild extension is a solution to the equations of general relativity, and it is a very interesting and beautiful one at that. It alerts us to the outrageous possibilities that spacetime may have in store for us and, as we shall see next, wormholes aren't even the half of it.

7

THE KERR WONDERLAND

In 1963, New Zealand mathematician Roy Kerr succeeded in finding the unique solution to Einstein's equations for a spinning black hole. Perhaps you might have expected that adding spin to Schwarzschild's 1916 solution should not be particularly taxing, but the fact that it took almost half a century to be achieved is testament to the complexity Kerr discovered. Kerr's solution, like Schwarzschild's, corresponds to an eternal black hole: an immortal twisting in empty space. But unlike Schwarzschild's, it is no longer spherically symmetric. Like most spinning objects, including the Sun and the Earth, a Kerr black hole bulges at the equator and is symmetric only about its axis of spin. This lack of symmetry has dramatic consequences.

There are two main types of Kerr black hole, which differ according to how fast they spin. We'll consider slowly spinning black holes first and get to the faster ones later. A slowly spinning Kerr black hole is illustrated in Figure 7.1. Compared to the Schwarzschild black hole, it has three new features. Firstly, the singularity is a ring.* The plane of the ring is aligned at a right-angle to the spin axis, which means that only trajectories in the

---

* The radius of the ring is J/c where J is the angular momentum of the hole divided by its mass and c is the speed of light. For the Earth J/c is roughly 10 metres and for the Sun it is roughly 1 kilometre.

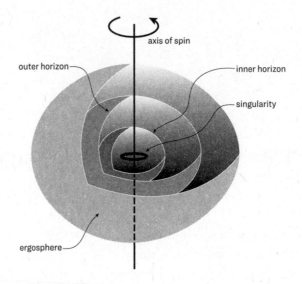

axis of spin

outer horizon

inner horizon

singularity

ergosphere

Figure 7.1. Schematic representation of a slowly spinning black hole.

equatorial plane will encounter it. All other trajectories will miss it. An astronaut could therefore fall into a Kerr black hole and dodge the end of time. Secondly, the hole has two event horizons, which we've labelled the inner and outer horizons. Thirdly, there is a region outside of the outermost horizon in which space is being dragged around so violently that it is impossible for anything to stand still.* This region is known as the ergosphere.

To appreciate the wonders of a spinning black hole, let's once again follow the adventures of an immortal astronaut. On descending towards the black hole, the first new feature our astronaut encounters is the ergosphere.† The outer surface of the ergosphere is the place where a light ray travelling radially outwards will freeze. In the Schwarzschild case, this is also the event horizon of the black hole: the place where the river of space

---

*   With respect to the distant stars, say.

†   'Encounter' is a bit misleading because the freely falling astronaut won't notice anything as she crosses into the ergosphere – as ever, spacetime for her will be locally flat.

is falling inwards at the speed of light, trapping the outward-swimming photon 'fish' forever. For the Kerr geometry, this place does not correspond to the event horizon – the place of no return. Our astronaut could pass into the ergosphere and then decide to turn around and escape back into the Universe. How can it be that a light ray heading radially outwards cannot escape, but an astronaut can? When the astronaut enters the ergosphere, it is impossible for her to avoid being swept around in the direction of rotation of the black hole. Space is being dragged around with the spin and no amount of rocketry can prevent the astronaut, or anything else, from being dragged around with it. This drag is the reason why an astronaut can beat a radially outgoing light ray and escape. We investigate this in more detail in Box 7.1.

BOX 7.1. A place where it is impossible to stand still

Figure 7.2 illustrates a rotating black hole, viewed along the axis of spin. Look at the little circles with nearby dots. The dots correspond to places where a flash of light is emitted, and the circles show the outgoing light front a few moments later. Far away from the hole, the dot lies pretty much at the centre of the circle, but as we move closer to the hole the dots become increasingly displaced. The circles are shifted inwards and also in the direction of the spin. Inside the ergosphere, the dots lie outside of the circles, and that is the important feature. For a Schwarzschild black hole, a similar thing happens inside the horizon: the dots lie outside of the circles, but in this case the circles are only pulled inwards. For a Kerr black hole, they are also 'dragged around' in the direction of the spin.

A way to picture this is to appreciate that, for a non-rotating black hole, a dot (flash) on the horizon produces an expanding spherical shell of light that must fall

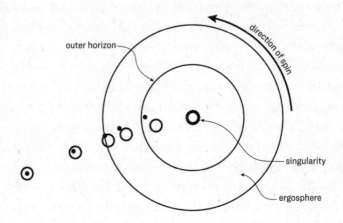

Figure 7.2. The ergosphere of a rotating black hole.

inwards because none of the light can ever travel beyond the
horizon. In the language of the river model, the light is being
swept inwards by the flow of space. For a rotating black hole,
the same is true, but there is also a swirling effect which
drags the circles around. It is possible for a dot to lie outside
of a circle, as shown in the diagram, which means that it is
not possible for someone who emitted the flash to stand still
at the position of the dot. If they did so, they would have
outrun the light they emitted. They therefore are forced to
swirl around with the black hole. This is the same idea we
discussed when considering observers falling into a
non-rotating black hole. In that case, observers cannot stand
still inside the horizon. In the rotating case, the same role
reversal is true in the ergosphere, which is a region from
which it is possible to escape.

We can see how it is possible to escape from the
ergosphere by focusing on the circle that straddles the
ergosphere (the third circle from the left). A little portion of
the circle lies outside the ergosphere. If we think of the circle
as the future light cone of the person who emitted the flash,

then we see that it is possible to draw a worldline for that person that crosses the boundary of the ergosphere and heads back outwards into the Universe beyond.

On crossing the ergosphere, our astronaut decides to continue inwards and cross the outer event horizon. As for the Schwarzschild black hole, this is a featureless place that marks the point of no return. The astronaut must now head inwards, and must cross the second, inner, event horizon. But there is a fascinating difference between the Schwarzschild and Kerr cases. For the spinning black hole, the astronaut regains her freedom to navigate once she has crossed the inner horizon. The singularity does not lie inexorably in her future, so time does not have to end. We can appreciate all of this by considering the spacetime diagram in Figure 7.3. Close to the outer event horizon, in region I, the geometry is similar to the Schwarzschild case. On crossing the outer horizon into region II, the light cones tip inwards, which means that the astronaut inexorably travels towards the inner horizon. However, on crossing the inner horizon into region III, the light cones tip back again and our astronaut can navigate around without encountering the singularity. Forever is still available. What, then, becomes of an astronaut who chooses to dodge the end of time?

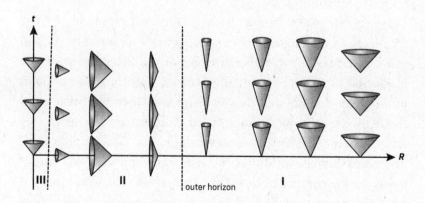

Figure 7.3. The future light cones outside and inside a Kerr black hole.

To answer this question, we need a Penrose diagram, which we can begin to construct given our experience with the Schwarzschild case. Figure 7.4 is a start. The purple line is the path of the astronaut as she travels from the universe outside (region I) through the outer horizon (the 45-degree black line) into region II and then through the inner horizon (the 45-degree orange line) into region III. Region I and the outer horizon are easy to draw because they are just like the Schwarzschild black hole. The two black lines on the right ($\Im^+$ and $\Im^-$) represent (lightlike) infinity and they are bona fide boundaries to the Penrose diagram. Our astronaut enters region II whence she is doomed to pass through the inner horizon, carried inexorably along by the flow of space. This mandates us to draw the orange lines representing the inner horizon. As for all horizons, they must be at 45 degrees. So far so good. Now comes the first novelty. The wiggly, vertical lines denote the ring singularity inside region III. Notice that they are vertical, and not horizontal as in the Schwarzschild black hole. This is because the Kerr singularity is timelike, which means our astronaut can see it (45-degree lines representing light rays starting from the singularity can reach the astronaut's worldline). This is different from the spacelike singularity inside a Schwarzschild black hole: the horizontal line on the Penrose diagram which nobody sees coming.

Given our experience with the Penrose diagrams in the last chapter, we immediately notice that Figure 7.4 cannot be the full story. The two upper diagonal edges of the diagram inside the black hole are horizons, not singularities, and they do not lie at infinity. We encountered a similar situation when we investigated the Schwarzschild black hole. It led us to extend the spacetime and discover the Einstein–Rosen bridge. The same applies here. To ensure that all worldlines end either at infinity or on a singularity, we are compelled to extend Figure 7.4. The result, part of which is shown in Figure 7.5, is quite shocking. We already know there can be an infinite volume of space inside the event horizon

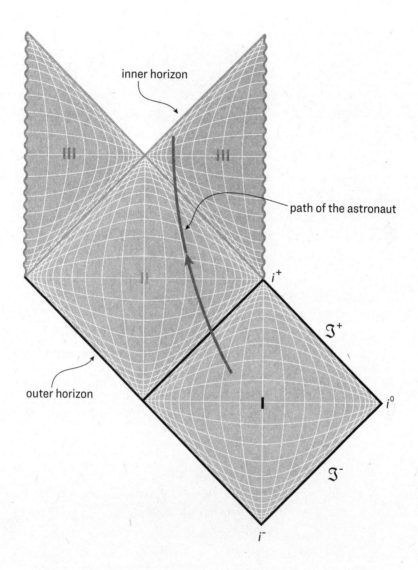

Figure 7.4. The Penrose diagram for a Kerr black hole corresponding to the discussion in the text. It is clearly incomplete. Also on plate 10.

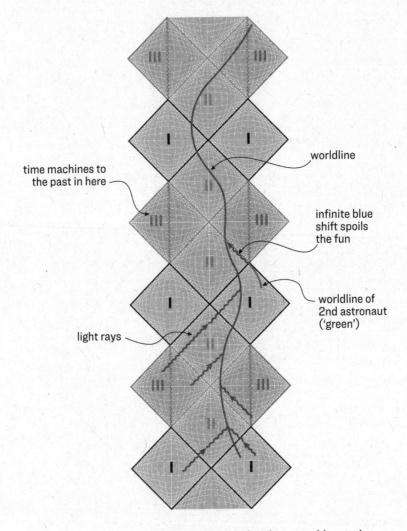

Figure 7.5. The maximal Kerr black hole. The vertical line is the worldline of our intrepid explorer. The part of her journey shown in Figure 7.4 is at the bottom. The wiggly lines represent possible paths of light rays. Also on plate 11.

of a black hole, but in the maximally extended Schwarzschild case we only had to contend with one extra infinite universe through the wormhole. Inside the eternal Kerr black hole there reside an infinity of infinite universes, nested inside each other like Russian TARDIS dolls. The Penrose diagram itself would fill an infinite sheet of paper. This Kerr wonderland is the unique way to extend Figure 7.4 and remain consistent with general relativity.

It would be entirely appropriate to ask 'What on earth have we drawn here?' The diagram depicts part of an infinite 'tower' of universes, which means there is an infinite amount of space and time tucked away inside an eternal Kerr black hole and no mandatory rendezvous with a singularity to spoil the fun. To see how things play out, let's resume the journey as our astronaut travels further into the interior of the black hole.

Recall, the astronaut started out in region I at the bottom of the diagram. This is the infinite region of spacetime outside the outer horizon. We might call it 'our Universe'. She crossed the outer horizon and entered region II. She is now inside the black hole, in between the outer and inner horizons. We can see that, just as for the Schwarzschild black hole, she will be able to receive signals not only from our Universe – region I – but also from another universe – the 'other' region I (we have drawn two wiggly light rays to illustrate that point). She could meet astronauts who have crossed the outer horizon from the other universe, but now they are not condemned to rendezvous with the singularity at the end of time because there is no singularity in region II. Instead, she must cross the inner horizon and enter region III. The ring singularity looms, denoted by the wavy vertical line, but she can avoid it. Now the fun starts.

She chooses to avoid the singularity and crosses into a second region II. This region is bounded by a horizon, but it is the horizon of a white hole, a gateway into a different region I – another universe. At this point, she might choose to head out and explore this new and enticing ocean of stars and galaxies, but she doesn't have to. There is another Kerr black hole outer horizon in this new universe, and she chooses to plunge across it. Once inside this second black hole, the whole story repeats itself until she dives into a third black hole. Emerging into region III at the top of the figure, she is now ready to brave the singularity. She dives through the ring and into another new universe in the infinite tower of universes. This universe is very different from the others, though. In this region of spacetime, gravity pushes rather than pulls.* It is an anti-gravity universe. It would still be possible for her to swing around and navigate back through the ring singularity. But it is also possible for her to arrange things so that she emerges before she entered it. This is possible because there are paths that our astronaut can take in region III which loop back and return to the same point. It's not possible to draw such paths on our Penrose diagram because they involve looping around in one of the dimensions we haven't drawn. These paths are known as 'closed timelike curves'. Imagine a path over spacetime that begins the day before your birth and, some years later (by your watch) arrives back at the day before your birth. This is time travel. Such paths are possible in the spacetime geometry in region III. This means that the Kerr black hole is a time machine (sometimes known as a Carter time machine, after Brandon Carter who first discovered it).

All sorts of issues now arise. What if the astronaut decided to prevent herself from being born? This isn't necessarily a paradox if she doesn't have free will; the universe could conceivably be constructed such that time travel is possible and yet the whole thing remains logically consistent. Perhaps it would be impossible for you to prevent yourself from being born and causing such paradoxes. Musing about free will is (perhaps inevitably, who knows) beyond the scope of this book, but asking questions about

---

* What our astronaut experiences in region III is not something we can appreciate from the Penrose diagram, but this is what the equations tell us.

possible spacetime geometries is not. The big question is: 'do spacetimes that permit closed timelike curves actually exist in Nature?'

In the proceedings of Kip Thorne's 60th birthday party – only the most eminent physicists have their birthdays written up in proceedings – Stephen Hawking writes about spacetimes that allow time machines.[22] 'This essay will be about time travel, which has become an interest of Kip Thorne's as he has become older ...,' he begins. 'But to openly speculate about time travel is tricky. If the press picked up that the government was funding research into time travel, there would be an outcry about the waste of public money ... So there are only a few of us who are foolhardy enough to work on a subject that is so politically incorrect, even in physics circles. We disguise what we do by using technical terms like "closed timelike curves" which is just code for time travel.'

Although it has not been proven beyond doubt, Hawking proposed a 'Chronology Protection Conjecture', which states that 'The laws of physics conspire to prevent time travel by macroscopic objects'. By macroscopic objects, Hawking means big things like astronauts rather than subatomic particles. The implication is that the maximally extended Kerr geometry should not exist in Nature, and we believe it does not for two reasons. Firstly, as we've already mentioned and will see in the following chapter, real black holes are made from collapsing matter. The presence of matter changes the spacetime inside the horizon of a black hole, effectively blocking up the portals into other universes. Both the maximally extended Schwarzschild and Kerr solutions are vacuum solutions to Einstein's equations – eternal black holes – and as far as we know, no such black holes exist.

The second reason why the Kerr wonderland shouldn't exist is illustrated in Figure 7.5. The short green curve is the worldline of someone moving without drama towards future timelike infinity in one of the region I universes. They send light signals to our astronaut inside the hole at regular intervals but, as we saw in

Chapter 3, there is an infinite amount of time compressed into the tip of the diamond. This means that the light signals pile up along the upper edge of the diamond and into the black hole along the inner horizon. This represents an infinite flux of energy (one of these signals is indicated by the purple wiggly line) which will result in the formation of a singularity, sealing off region III, the ring singularity and beyond. 'The inner horizon marks the last moment at which our astronaut can still receive news, but then she gets all of the news.'[23] Worldlines will end on the singularity* so no extension is necessary or possible into the region containing the ring singularity, time machines and the infinite tower of universes.

Fast-spinning black holes

If the spin (J/c) of the hole is bigger than one half of the Schwarzschild radius, the Penrose diagram is not that of the Kerr wonderland. The spacetime is vastly simpler, but involves what is known as a naked singularity, as illustrated in Figure 7.6.

The event horizons have disappeared, leaving just a ring singularity (the wiggly line), which remains a portal to an infinite space where gravity repels instead of attracts. A naked singularity is a singularity from which the Universe is not protected by an event horizon. Naked singularities are an anathema to physicists. So much so that Roger Penrose was moved to introduce the 'cosmic censorship conjecture', which asserts that no naked singularities exist in our Universe other than at the Big Bang. The trouble with naked singularities is that they contaminate spacetime with ignorance; the world becomes hopelessly non-deterministic. Perfect knowledge of the past would be insufficient to predict the future. To see why, imagine any event in the spacetime of Figure 7.6. There will be light rays that can reach this event, and therefore

---

* A horizontal, spacelike singularity like the one in a Schwarzschild black hole.

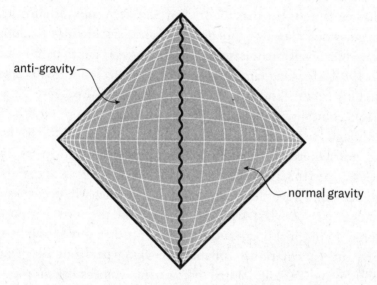

Figure 7.6. The Penrose diagram for a fast-spinning black hole.

influence it, coming from the singularity. The singularity, however, is a place where the known laws of physics break down. This means that every event in the spacetime can be influenced by an unpredictable region of spacetime, and this is a nasty situation for physicists who are in the business of predicting the future from a knowledge of the past. Having said that, Nature is not obliged to make physicists' lives easier.

In 1991, Kip Thorne, John Preskill and Stephen Hawking made a famous bet that the laws of physics would forbid naked singularities. Hawking thought that naked singularities would be forbidden in all circumstances, but this appears not to be the case. In 1997, Hawking famously conceded the bet 'on a technicality' and his concession made the front page of the *New York Times*. The technicality is that computer simulations can produce them, although the models are highly contrived. Having said that, they do not require anything too exotic beyond the known laws of physics. Thorne, Preskill and Hawking therefore renewed their bet with modified wording. No naked singularities will occur *naturally* in our Universe without the need for the intervention of some unimaginably advanced civilisation that could arrange to fine tune gravitational collapse. Hawking paid out by giving Thorne and Preskill t-shirts 'to cover the winner's nakedness', in accord with the wording of the bet. The t-shirts were so politically incorrect, in Kip Thorne's words, that they were forbidden from ever wearing them.

What prevents a Kerr black hole naturally acquiring sufficient spin to produce a naked singularity? One could easily imagine making a black hole spin faster, such that even though it started out spinning slowly, it ends up spinning fast enough to expose a naked singularity. For example, why not drop a spinning ball (or maybe a star if we want to get serious) into the hole and arrange things so that the spin is in the same direction as the spin of the hole. That would increase the hole's spin, potentially pushing it over the critical value. This calculation can be done in general relativity, and it turns out that the hole pushes the spinning thing away. This 'spin-spin' interaction is a nice example that illustrates how the theory of general relativity appears to be constructed such that cosmic censorship holds. It seems that naturally occurring black holes always have their singularities tucked away behind horizons.

You may be disappointed that Nature does not appear to permit wormholes and Kerr wonderlands to exist, but your disappointment may be too pessimistic. The message is that general relativity has a richness that permits a remarkable gamut of spacetimes. Perhaps some of this extraordinary potential is realised in Nature? We will return to this cryptic statement later: the answer isn't a straight 'No'.

Return to the ergosphere

While the interior geometry of the Kerr black hole may be eliminated by the in-falling matter of the collapsing star that formed it, this is not true of the ergosphere which sits beyond the outer horizon. Spinning black holes do exist, and the external spacetime is described by the Kerr solution. So let us return to the ergosphere; the region just beyond the outer horizon within which it is impossible not to get swept around by the flow of space. Recall that, inside the ergosphere, space and time swap roles (see Box 7.1), but it remains possible to escape. Roger Penrose first appreciated the consequences of this space and time role reversal inside the ergosphere; it makes it possible to extract energy from a rotating black hole. The idea is illustrated in Figure 7.7.

Imagine throwing an object into the ergosphere, where it breaks into two pieces. One piece falls into the black hole and the other piece comes back out. This is possible, because the ergosphere lies outside of the event horizon. The surprising thing is that the piece that comes out can carry more energy than the original object carried in. How does this magic come about?

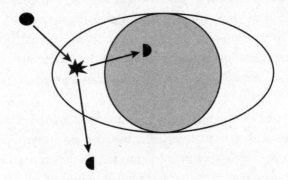

Figure 7.7. The Penrose process.

The important insight is that inside the ergosphere objects can have negative energy from the perspective of someone outside the hole. Outside of a black hole, objects always have positive energy. Inside the ergosphere, however, it becomes possible for negative energy objects to exist.* This possibility arises because of the swapping of the roles of space and time and because energy and momentum are intimately related to space and time.

To appreciate the link between space, time, momentum and energy, we need a brief detour back to 1918 and to Amalie Emmy Noether. Noether was, in Einstein's words, 'the most significant creative mathematical genius thus far produced since the higher education of women began'. Among her many achievements, Noether discovered that the law of conservation of energy is a direct consequence of time translational invariance, which when the jargon is stripped away means that the result of an experiment does not depend on what day of the week it is performed (all other things being equal). Likewise, the law of momentum conservation is a consequence of translational invariance in space, which means the result does not depend on where the experiment is performed (all other things being equal). This means that the

---

* Using notions of energy and time defined by an observer far from the hole.

role reversal of space and time in the ergosphere is accompanied by a corresponding reversal of the roles of momentum and energy. In the Universe outside the black hole – our everyday world – momentum can be positive or negative because things can move left or right. Inside the ergosphere, the switch means that energy can likewise be positive or negative.

If a piece of the in-falling object breaks away inside the ergosphere and carries negative energy into the black hole, the energy of the black hole will decrease. Energy must be conserved overall, however, and this means the outgoing part must carry away more energy than was thrown in.

In Figure 7.8 we reproduce a figure from Misner, Thorne and Wheeler showing how an advanced civilisation living around a

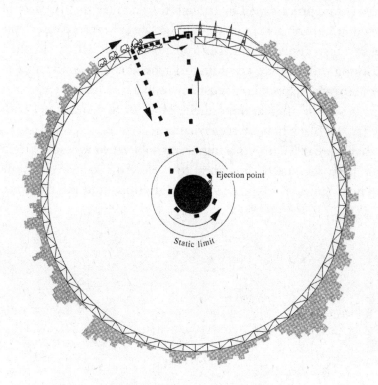

Figure 7.8. Black hole mining. From Misner, Thorne and Wheeler's 1973 book, *Gravitation*.

Kerr black hole could exploit the Penrose process to dispose of their garbage and generate electrical power for their civilisation. The ultimate green energy scheme.

We've spent quite some time exploring the ergosphere because there's an important postscript. It transpires that the area of the black hole's event horizon always increases after a Penrose process. At first sight this is surprising because a rotating black hole loses mass through the Penrose process. We might therefore expect the horizon to shrink. We are considering rotating black holes, however, and the area of the outer event horizon can increase even if the mass of the black hole decreases, provided that the spin of the black hole also decreases. Using the equations of general relativity, it can be shown that the spin of a black hole always decreases in a Penrose process, and by enough to guarantee that the area of the outer horizon always increases. This 'area always increases' rule doesn't just apply to the Penrose process. In 1971, Stephen Hawking proved that, according to general relativity, the area of the horizon of a black hole must *always* increase, no matter what.* This is a result of great importance. It is our first encounter with the laws of black hole thermodynamics.

Before we delve into this important subject, we will take a step back from the purely theoretical to explore the formation of the real black holes we observe to be dotted throughout the Universe.

---

* It will be allowed to decrease when we come to consider quantum physics.

8

REAL BLACK HOLES FROM COLLAPSING STARS

'In my entire scientific life, extending over forty-five years, the most shattering experience has been the realization that an exact solution of Einstein's equations of general relativity, discovered by the New Zealand mathematician, Roy Kerr, provides the absolutely exact representation of untold numbers of massive black holes that populate the universe. This shuddering before the beautiful, this incredible fact that a discovery motivated by a search after the beautiful in mathematics should find its exact replica in Nature, persuades me to say that beauty is that to which the human mind responds at its deepest and most profound.'

Subrahmanyan Chandrasekhar[24]

The black holes we've explored so far have inhabited the mathematical landscape of general relativity. These remarkable universes were known to and broadly speaking dismissed by physicists for a large part of the twentieth century, Einstein included, on the very reasonable grounds that we shouldn't conclude that something exists just because a physical theory allows it. If black holes are to exist in the real sky rather than the mathematical one, Nature must construct them. Real black holes, formed from stellar collapse, are the focus of this chapter. We will learn that the solutions to Einstein's equations of general relativity discovered by Schwarzschild and Kerr are of extraordinary significance in the real Universe because they are the only possible solutions for the spacetime in the region outside of every black hole. Nowhere else in physics is something apparently so complicated as a collapsing star reduced to something so simple and with such precision. The Schwarzschild solution depends on just one number (the mass) and the Kerr solution adds a second number (the spin). Knowing these two numbers alone, we can compute the gravitational landscape in the region outside of real black holes, *exactly*. That is an astonishing claim – it does not matter what collapsed to form the black hole, nor does it depend on how it fell in. All that remains outside the horizon is a spacetime perfect in its simplicity. This is what moved Chandrasekhar to write the powerful prose quoted above. Quoting Chandrasekhar again: 'The black holes of nature are the most perfect macroscopic objects there are in the universe … and since the general theory of relativity provides only a single unique family of solutions for their descriptions, they are the simplest objects as well.'

John Wheeler, as ever, found a more pithy phrasing: 'Black holes have no hair.' In his memoir *Geons, Black Holes and Quantum Foam*, Wheeler recounts an exchange with Richard Feynman in which the often irreverent Feynman accused him of using language 'unfit for polite company'. 'I tried to summarize the remarkable simplicity of a black hole by saying a black hole has no hair. I guess Dick Feynman and I had different images in mind. I was thinking of a room full of bald-pated people who were hard to identify individually because they showed no differences in hair length, style, or colour. The black hole, as it turned out, shows only three characteristics to the outside world: its mass, its electric charge (if any) and its spin (if any). It lacks the "hair" that more conventional objects possess that give them their individuality … No hair stylist can arrange for a black hole to have a certain colour or shape. It is bald.'

A series of papers throughout the late 1960s and early 1970s established the magnificent simplicity of black holes, as viewed from the outside. Once formed, according to general relativity, the horizon shields us from the complexities within. Even if something as large as a planet or star falls across the horizon, the American physicist Richard H. Price proved, in 1972, that the black hole quickly settles down again into oblivious perfection. For a Schwarzschild black hole, the horizon will reassume the shape of a perfect sphere, and any disturbances caused by the in-falling body will be smoothed out by the emission of gravitational waves. The conclusion is that the spacetime outside of all black holes in the Universe is either Schwarzschild or Kerr.*

What, then, happens in the case of a real collapsing star? Is it possible or even inevitable that a dense enough lump of matter will fall inwards to create a horizon and ultimately disappear into a spacetime singularity? The first attempt at addressing this question came back in 1939, when Robert Oppenheimer and Hartland Snyder showed that a star will collapse to form a black hole under certain assumptions. Specifically, they considered a pressureless ball of matter with perfect spherical symmetry. You may well baulk at this: the interior of a star is certainly not a zero-pressure environment, and the collapsing matter isn't a perfect sphere. Perhaps the Oppenheimer–Snyder conclusion that black holes can form in Nature is associated with the assumption of perfect spherical symmetry. If everything is falling towards a single, precise point in the middle of the ball then no wonder something weird happens. The more realistic case will have matter swirling around and involve all the complexity of real stars. Maybe that leads to a collapse that does not generate a spacetime singularity. For many years, the possibility that black holes do not form out of collapsing stellar matter remained a mainstream view.

---

* Strictly speaking, black holes can also carry electric charge but astrophysical black holes are electrically neutral.

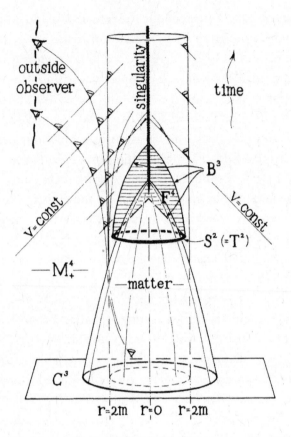

Figure 8.1. The spacetime diagram for a collapsing star, from Penrose's
1965 paper 'Gravitational Collapse and Space-Time Singularities'.

The publication of Roger Penrose's paper in January 1965
essentially resolved the issue. He showed that the complex dynamics of the stellar collapse does not matter, and black holes must
form if certain conditions are met.*

Penrose demonstrated that the formation of a spacetime singularity is inevitable once a distribution of matter has become so

---

* The waters were muddied by a Lifshitz and Khalatnikov paper of 1963 which
  claimed to prove no singularity would occur. Following work with Belinskii, Lifshitz
  and Khalatnikov withdrew their claim in 1970.

compressed that light cannot escape from it. Figure 8.1 is taken from Penrose's paper (it is hand-drawn by Penrose himself) and provides an intuitive way of picturing the collapse of a star to form a black hole. Time (as measured by someone far away from the star, labelled 'outside observer' on the diagram) runs from the bottom to the top, and one of the three space dimensions is not drawn. The surface of the star is therefore drawn as a circle on any horizontal slice through the diagram. For example, on the slice labelled C^3 at the base of the diagram the star's surface is represented by the solid black circle. The dotted circle inside represents the Schwarzschild radius of the star* (recall that for the Sun the Schwarzschild radius is 3 kilometres). All the complex physics of the stellar interior plays out inside the solid circle and the beauty of Penrose's argument is that the details of what's happening there do not matter once the star has collapsed inside its Schwarzschild radius.

We can follow the collapse of the star by moving upwards on the diagram. Each horizontal slice corresponds to a moment in time. As time passes, the circles representing the surface of the star get progressively smaller and the stellar surface traces out the cone-like shape on the diagram. The interior of the cone is the interior of the collapsing star and is labelled 'matter'. The vertical dotted lines mark out the Schwarzschild radius. They become solid lines when the black hole has formed and then denote the event horizon. There is a lot of detail on this diagram that we don't need – it is after all taken directly from Penrose's published paper. It is worth looking at the light cones, however. Once the stellar surface has passed through the Schwarzschild radius the light cones inside all point towards the singularity. The outside observer therefore never sees the star collapse through the horizon. They see an increasingly slow contraction of the star as its surface

---

*   On the diagram, the Schwarzschild radius is at $r = 2m$ where m is the mass of the star because Penrose is working in units where $G = c = 1$.

approaches the horizon. For anyone inside the horizon it's easy to see that the singularity lies inexorably in their future, although they never see it coming. Again, we see that the singularity is a moment in time.

While Penrose drew his diagram for the Schwarzschild case of zero spin, his theorem is more general and also applies to Kerr black holes, or to any conceivable collapsing distribution of matter. The theorem is concerned with what is happening at the solid black circle labelled S^2, which forms before the singularity. This imaginary surface is known as a 'trapped surface', and it's the key element in Penrose's argument because he demonstrated that not all light rays will continue to propagate forever if the space-time contains a 'trapped surface'. So what is this trapped surface?

Figure 8.2 illustrates the idea. Picture some blob-like region of space and imagine lots of pulses of light flashing from the surface of the blob. For a blob in ordinary flat space, half of the light will head outwards, away from the blob, and the other half will head inwards. That is illustrated on the left of the figure. We've only shown five flashes of light, but we imagine many more. The black wavy lines represent light heading out and the grey wavy lines

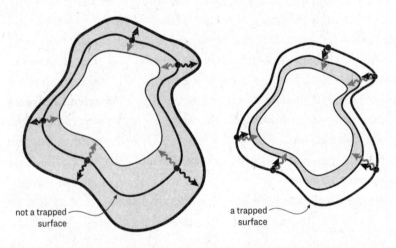

Figure 8.2. A trapped surface.

represent light heading inwards. The shaded region is the volume between these two sets of flashes and it will grow with time as the flashes head outwards and inwards at the speed of light. Since nothing travels faster than light, any matter initially sitting on the surface of the blob must stay in the expanding, shaded region. So far so good (hopefully).

On the right we've drawn a trapped surface. In this case, both the grey and black flashes are heading inwards. This happens inside the horizon of a black hole due to the curved geometry of spacetime. The converging of the light rays spells trouble. As before, any matter sitting on the trapped surface must stay inside the shaded region because nothing can travel faster than light. But now this region is shrinking down to nothing. In Penrose's diagram the shaded region labelled F^4 corresponds to the shaded region in Figure 8.2.

You might suppose that this is obvious since all matter inside the trapped surface is destined to get squeezed down to nothing, but we should be careful when wielding our dodgy intuition like this. As we've learned in the case of the Kerr black hole, matter might slip through a wormhole to explode into an infinite space-time 'on the other side'. What Penrose demonstrated rigorously is that at least one in-falling light ray will terminate. The mathematical techniques Penrose employed in his 1965 paper opened the door to a series of successively more wide-ranging singularity theorems, developed mainly by Penrose in collaboration with Stephen Hawking. Significantly, they managed to extend Penrose's original theorem to include all particles (not just rays of light). They also applied the theorems 'in reverse' to show that in general relativity the Universe must have a singularity in the past which, to repeat the quote from the beginning of this book, '... constitutes, in some sense, a beginning to the universe'.

As something of an aside, it's notable that the singularity theorems alone do not guarantee that a black hole will form in all circumstances. Black holes are not just singularities; they are black holes because their interior is shielded from their exterior by an event horizon. As we've seen, there could conceivably be singularities that are not shielded by a horizon such as the naked singularity in a fast-spinning Kerr black hole. To avoid that possibility, we also need the cosmic censorship conjecture as discussed in Chapter 7.

Naked singularities aside, the only way to avoid the conclusion that black holes must exist in our Universe is to argue that it's not possible for matter to be squashed down sufficiently to form a trapped surface. That does not seem likely since we know from the work of Chandrasekhar that no known physics can halt the collapse of sufficiently massive stars. One might try to argue that some dramatic and unanticipated astrophysics or some new force of Nature steps in to halt the collapse before a trapped surface forms. Perhaps sufficient matter gets blown away as the gases swirl or as the collapsing star implodes. That might happen, but it is unlikely to be the case for every possible collapsing system. To emphatically illustrate the point, Penrose's theorem applies if a large number of ordinary stars are close enough to form a trapped surface around them, such as could conceivably happen in the centre of a galaxy. In that case, the stars could still be very far apart so that the average density of matter is far less than the average density of a star, and we understand physics at these densities very well. Nevertheless, the theorem tells us that the stars are doomed to collapse.

The singularity theorems of Penrose and Hawking marked a change in the way physicists regarded black holes: combined with the work of Chandrasekhar, the theorems served to convince virtually all physicists that, in the words of the Nobel Prize committee, 'black holes are a robust prediction of general relativity'. Today, we don't need to rely on the theory alone because our twenty-first-century technology has allowed us to take photographs of supermassive black holes and to observe black hole collisions using gravitational wave detectors.

The Penrose diagram of a real black hole

When we encountered the wormholes and wonderlands of the eternal Schwarzschild and Kerr solutions to Einstein's equations, we said that the portals to other universes were located in a part of the Penrose diagrams that would not exist inside a black hole formed by the gravitational collapse of a star. What, then, does the Penrose diagram of a real astrophysical black hole look like?

In 1923, the American mathematician George David Birkhoff proved that the spacetime outside of any spherical, non-rotating distribution of matter must be the Schwarzschild spacetime. This remains true even if the matter is in the process of collapsing. With this extra piece of information, we can draw the Penrose diagram corresponding to an entire spacetime in which a lone black hole forms out of a collapsing, spherical shell of matter.

Imagine a thin spherical shell of matter in the process of collapsing. A real star will of course not be just a shell, so we are simplifying a little. Outside of the shell, according to Birkhoff, the spacetime will be that of Schwarzschild. Inside the shell, there is no gravity. This is also true in Newton's theory of gravitation, as Newton himself proved in his *Principia Mathematica*. Inside, therefore, the spacetime is flat. All we need to do to make the Penrose diagram therefore is to stitch together two bits of spacetime corresponding to the interior and exterior of our shell. We do that in Figure 8.3.

The top left diagram is flat (Minkowski) spacetime as depicted in Figure 3.10. The blue shaded region is the region inside of our collapsing shell. The curving black line is the worldline of the shell. We assume it starts collapsing in the distant past (the bottom of the triangle, which corresponds to past timelike infinity) and that it shrinks to zero radius at some finite time. The interior of the shell, i.e. the blue shaded region, is Minkowski spacetime, because there is no gravity there. The exterior of the shell (the unshaded region) is, according to Birkhoff's theorem,

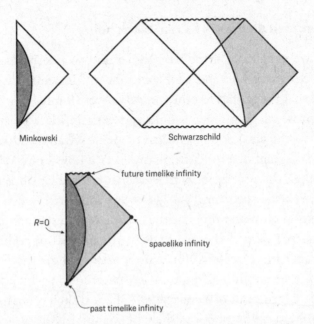

Figure 8.3. The Penrose diagram corresponding to a collapsing shell of matter (bottom). The inside of the shell is Minkowski spacetime (blue on plate 12; here it's the darker grey) and the outside of the shell is Schwarzschild spacetime (red on plate 12; here it's the lighter grey). Also on plate 12.

Schwarzschild spacetime. That is illustrated in the top right diagram by the shaded red region. The curving line is the world-line of the collapsing shell again, but this time drawn in Schwarzschild spacetime. The complete spacetime must be Minkowski inside the shell and Schwarzschild outside, which means it must look like the lower diagram in the figure, which is obtained by patching together the blue and red portions of the upper diagrams.* The disappointing result is that there is no wormhole or white hole anymore – the region inside the shell is 'boring-flat-old' Minkowski spacetime.

---

* Strictly speaking this logic works only for the region outside of the horizon and we are guessing somewhat as to what happens inside the horizon.

We can get a different picture of what's happening during the collapse of the star by drawing some embedding diagrams, just as we did for the wormhole inside the eternal Schwarzschild black hole. In Figure 8.4 we show a series of embedding diagrams which represent slices through the Penrose diagram at progressively later times as the shell collapses. Time runs from top left to bottom right. Initially, the shell is very large and not particularly dense and barely makes a dint in the otherwise flat spacetime. We've also shown an astronaut named 'A' who decides to follow the collapsing shell inwards. The astronaut sees the shrinking shell below them. The second row shows the situation at some time later. The shell is now smaller and denser. When its radius shrinks inside the Schwarzschild radius, a black hole is formed. The astronaut, unbeknown to them, has also passed through the event horizon, but they don't notice anything out of the ordinary. They still see the shell below them. The lower diagram is close to the moment of the singularity: the superdense shell has distorted space dramatically. Even though A is very close to the singularity, they still see the shell way down below. In a sense the shell is blocking up the wormhole that would have been present in the eternal Schwarzschild geometry. The singularity is the moment when the space gets infinitely stretched and infinitely thin, at which point the astronaut and the shell cease to exist.

Horizons: shuddering before the beautiful

Black holes, then, exist in our Universe and we are forced to confront their intellectual challenges. As Chandrasekhar wrote, 'this incredible fact that a discovery motivated by a search after the beautiful in mathematics should find its exact replica in Nature, persuades me to say that beauty is that to which the human mind responds at its deepest and most profound'. And yet there is something at first sight that is deeply disappointing about the way

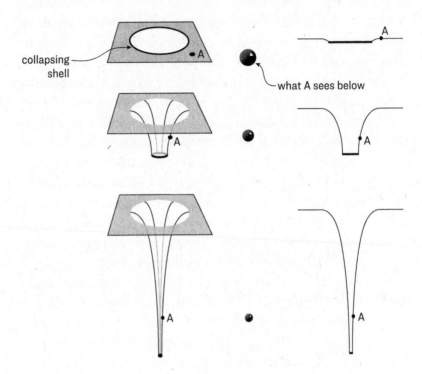

Figure 8.4. A collapsing shell of matter is one way to make a black hole. The shell shrinks and increasingly distorts the spacetime nearby. The side view on the right helps us to see the curving of space. Astronaut A falls in behind the shell and always sees it way below them, shrinking as it recedes due to tidal effects. The ball in the middle column is what they would see in three dimensions. The singularity is the moment in time when the 'throat' becomes infinitely long and infinitely narrow. The third row corresponds to a spatial slice just before the singularity. You can see that the shell is still way down below the astronaut. Neither shell nor astronaut actually hit anything as they fall to their doom. Rather, the singularity is a 'pinching to nothing' of the space at a moment in time.

Nature has chosen to realise this beauty, because it appears that the real treasure will be forever hidden from us. Black holes have horizons, and the horizon would appear to ensure that we must forever remain blind to the details of the collapsing material that falls into the singularity. That blindness is what gives rise to the remarkable applicability of the exterior Kerr and Schwarzschild solutions that Chandrasekhar refers to. Every black hole is identical, save for its spin and mass. Black holes have no hair. On the one hand this is a beautiful thing, but on the other hand it is bad news because it means we cannot observe the singularity to learn more. Cosmic censorship is highly desirable if we wish to maintain predictability, because we have no idea what the laws of physics are at the location of the singularity. By stashing the singularity behind a horizon, Nature appears to protect physicists living outside of a black hole from their ignorance of the singularity. But physicists don't want to be protected. We want to learn about what happens when the known laws break down and must be replaced by the holy grail of theoretical physics – a quantum theory of gravity. To date, there is no proof of cosmic censorship, but if there is such a censor, we may never have direct access to the clues we need to explore quantum gravity.

This was the view of many theoretical physicists until very recently. In the last few years, however, there has been a realisation that the clues to quantum gravity may not only reside near the singularity. They may also be found in the physics of the horizon. This came as a very welcome surprise, because it was long assumed that whatever happens close to the event horizon of a black hole, the physical processes should have nothing at all to do with the extreme conditions at the singularity where we expect quantum gravitational effects to be important. The horizon, after all, is a place in space through which an astronaut can happily fall and suffer no ill effects. This assumption now appears to have been too pessimistic. The study of the thermodynamics of black holes, which began in the 1970s as an investigation into quantum mechanical effects in the vicinity of the horizon, has opened an entirely unexpected window into the deep mystery of quantum gravity. And it is to black hole thermodynamics that we now turn.

9

BLACK HOLE THERMODYNAMICS

'Black holes ain't so black'

Stephen Hawking

Up to this point, we've thought about black holes as objects that, broadly speaking, mind their own business. Stuff can fall in, causing the black hole to grow, but nothing that crosses the horizon can come out. All traces of anything and everything that falls into a black hole would seem to be erased from the Universe forever. This is the description of a black hole according to general relativity. In 1972, John Wheeler and his graduate student Jacob Bekenstein realised that this raises a deep question. Wheeler tells the story of how he, in a 'joking mood one day', told Bekenstein that he always felt like a criminal when he put a cup of hot tea next to a glass of iced tea and let them come to the same temperature. The energy of the world doesn't change, but the disorder of the Universe would be increased, and that crime 'echoes down to the end of time'.[25] Wheeler was referring to the Second Law of Thermodynamics, which roughly speaking says that when any change occurs in the world, the world becomes more disordered as a result. 'But let a black hole swim by and let me drop the hot tea and the cold tea into it. Then is not all evidence of my crime erased forever?' Jacob took the remark seriously, recounts Wheeler, and went away to think about it.

The physical quantity that measures disorder is known as entropy. In Wheeler's words, 'Whatever is composed of the fewest number of units arranged in the most orderly way (a single, cold molecule for instance) has the least entropy. Something large, complex and disorderly (a child's bedroom perhaps) has a large entropy.' The Second Law of Thermodynamics, phrased in terms of entropy, states that in any physical process, entropy always increases. Wheeler was concerned that he had increased the entropy of the Universe by placing his two cups of tea in contact, and then decreased it again by throwing them into a black hole.

As is so often the case, there is a lyrical Eddington quote that reflects the importance of the Second Law: 'The law that entropy always increases holds, I think, the supreme position among the laws of Nature. If someone points out to you that your pet theory of the universe is in disagreement with Maxwell's equations – then so much the worse for Maxwell's equations. If it is found to be contradicted by observation – well, these experimentalists do bungle things sometimes. But if your theory is found to be against the Second Law of Thermodynamics I can give you no hope; there is nothing for it but to collapse in deepest humiliation.'

Bekenstein returned a few months later with an answer: the black hole does not conceal the crime. His answer was inspired by Hawking's prior observation that the area of a black hole's horizon always increases, no matter what. To Bekenstein, this 'area always increases' law reminded him of the 'entropy always increases' law. He therefore made the bold claim that throwing objects into a black hole causes an increase in the area of the event horizon, which in turn signals a corresponding increase in entropy. In other words, when something falls into a black hole, a record is kept, the Second Law is obeyed, and deepest humiliation is avoided. Looking back at Wheeler's description of entropy in terms of molecules and messy bedrooms, however, the assignment of an entropy to a black hole would seem to be questionable, as Bekenstein and Wheeler were well aware. According to general relativity, black holes are simple things: a Schwarzschild black hole can be described by a single number; its mass. But Wheeler begins his description of entropy with the phrase 'Whatever is composed of the fewest number of units arranged in the most orderly way ... has the least entropy.' With no apparent units to re-arrange, what, then, is the meaning of the entropy of a black hole?

Entropy was introduced in the nineteenth century as one of the fundamental quantities in the newly emerging science of thermodynamics, alongside the more familiar notions of heat, energy and temperature. Perhaps because of its historical origins in the industrial revolution, thermodynamics has the reputation of being closer to engineering than black holes, but this is emphatically not the case. Thermodynamics is connected at a deep level to quantum mechanics and the structure of matter. When applied to black holes, we will see that thermodynamics is connected at a deep level to quantum gravity and the structure of spacetime. Before we get to the thermodynamics of black holes, let's go back in time to the nineteenth century for a tour of the origins of the subject and to introduce the important concepts of heat, energy, temperature and entropy.

The fundamental physics of the fridge

The foundations of thermodynamics were laid by practical people doing practical things: people like Salford brewer James Prescott Joule who were interested in building better steam engines, developing more efficient industrial processes, and beer.

In the early 1840s, Joule performed a range of experiments to demonstrate that heat and work are different but interchangeable forms of energy. Figure 9.1 illustrates his most famous experiment. A weight falling under gravity turns a paddle that stirs some water, causing the temperature of the water to rise. In thermodynamical jargon, the falling weight does work on the water. Joule's skill was in being able to make very precise measurements of the temperature increase, which he demonstrated to be proportional to the amount of work done by the falling weight. Initially, his findings were not met with enthusiasm. They went against the thinking of the day, which held that heat is an ethereal fluid ('caloric') that flows from hot to cold objects. Joule submitted his findings to the Royal Society in 1844, but his paper was rejected, partly because it was not believed that he could measure temperature increases to the claimed 1/200th of a degree Fahrenheit. The Royal Society, then as now, was not awash with experts in the brewing of fine ales, the demands of which meant that Joule had access to instruments capable of the required precision. A notable exception is the former President of the Royal Society and Nobel Laureate Sir Paul Nurse, who began his distinguished career as a technician in a brewery because he wasn't accepted onto a university degree course due to his lack of a modern language qualification. Sir Paul was subsequently awarded the 2001 Nobel Prize for his work on yeast. Joule remained undaunted, and by the mid-1850s his work had become widely accepted after a fruitful collaboration with William Thomson (later Lord Kelvin).

Joule's results illustrate what we now know to be correct: heat is a form of energy associated with the motion of atoms and molecules – the building blocks of matter. As the paddle rotates, it delivers kinetic energy to the water molecules by hitting them. The molecules move around faster, and this is what we measure as an increase in the temperature of the water. At the time, this idea was radical because there was no direct evidence that matter is composed of atoms, although Joule was taught by one of the leading proponents of the atomic hypothesis, John Dalton. In the words of Jacob Abbott, writing about Joule's experiments in 1869:

'It is inferred from this that heat consists in some kind of subtle motion – undulatory, vibratory or gyratory – of the elemental atoms or molecules of which all material substances are supposed to be composed. This, however, is a mere theoretical inference.'[26]

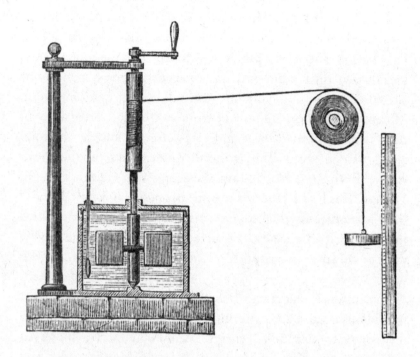

Figure 9.1. James Prescott Joule measured the increase in temperature of a vessel of water caused by the rotation of a paddle driven by a weight falling under gravity. The experiment demonstrated the conversion of mechanical work into heat.

The link between work, temperature, and the motion of the proposed atomic constituents of matter is a clue that thermodynamics is related to the behaviour of the hidden building blocks of the world, whatever those building blocks may be. Incidentally, one of the papers that settled the atomic debate was Einstein's 1905 paper on Brownian motion (and his follow-up in 1908), which explained the jiggling of pollen grains suspended in water under the assumption that they were being bombarded by water molecules. Einstein's predictions were confirmed experimentally in 1908 by Jean Baptiste Perrin, who received the Nobel Prize in 1926 for his work on 'the discontinuous structure of matter'.

The results of Joule's experiments, as well as providing evidence for the existence of atoms, are captured in what we now call the First Law of Thermodynamics, which expresses the fundamental idea that energy is conserved: The total energy of a system can be altered either by supplying or extracting heat or by doing work. Moreover, a certain amount of work can be converted into an equivalent amount of heat and vice versa, so long as the total energy is conserved. This is the theoretical basis of the steam engine. Burn some coal and use the energy that's released to spin a wheel. This isn't all there is to a steam engine, however, because there is another essential component – the environment in which the engine sits. Crucially, the surroundings of the steam engine must be colder than the furnace, otherwise the steam engine won't work. Why?

The answer is that energy is always transferred from hot objects to cold objects and never the other way round. This has nothing to do with the conservation of energy. Energy would still be conserved if it was removed from a cold cup of tea, making it colder, and transferred to a hot cup of tea, making it hotter. But this is not what happens in Nature. To account for this one-way transfer of energy, another law of Nature is required, and that is the Second Law of Thermodynamics. One way to state the Second Law is simply to say that heat always flows from hot to cold. Described in these terms, a steam engine is a device that sits between a hot furnace and the cold world outside. As energy flows naturally from hot to cold, the engine syphons off some of the flow and converts it into useful work. Hardly of profound significance at first sight, but it turns out that this almost self-evident statement of the Second Law captures the essence of a much deeper idea. In his book *The Laws of Thermodynamics*, Peter Atkins begins his chapter on the Second Law with this remarkable sentence: 'When I gave lectures on thermodynamics to an undergraduate chemistry audience I often began by saying that no other scientific law has contributed more to the liberation of the human spirit than the second law of thermodynamics.' 'The second law,' he continues, 'is of central importance in the whole of science, and hence our rational understanding of the universe, because it provides a foundation for understanding why *any* change occurs. Thus, not only is it a basis for understanding why engines run and chemical reactions occur, but it is also a foundation for understanding those most exquisite consequences of chemical reactions, the acts of literary, artistic, and musical creativity that enhance our culture.'[27]

German physicist Rudolph Clausius introduced the idea of entropy in 1865. In his words: 'The energy of the world is constant. The entropy of the world strives for a maximum.'* This is a beautifully succinct statement of the first two laws of thermodynamics. Let's see how things work out in the case of Wheeler's teacups. According to the First Law, energy is always conserved. This can be true whichever way the energy flows, as long as the amount of energy removed from one teacup is equal to the amount of energy deposited in the other. Clausius defined entropy such that the entropy increase caused by adding heat energy to cold tea is greater than the entropy decrease when the same amount of energy is removed from hot tea.† Thus, the combined entropy of the two cups will increase if heat flows from hot to cold, but not the other way round.

Energy *can* flow from a cooler object to a hotter object if, somewhere else, enough energy is dumped into another cooler object such that the overall entropy of everything increases. This is what happens in a fridge. Heat is removed from the inside, which lowers the entropy of the interior. The entropy of your kitchen must therefore increase by a larger amount to satisfy the Second Law. This is why the back of your fridge must be hotter than your kitchen. Here's how it works.

---

* Taken from Clausius's excellently titled 1865 paper, 'The Main Equations of the Mechanical Heat Theory in Various Forms that are Convenient for Use'.

† Specifically, the change in entropy $dS = dQ/T$, where dQ is the amount of heat energy transferred to the cup of tea and T is its temperature.

A coolant circulates around the fridge from the interior to the exterior. On exiting the interior, the coolant is compressed and therefore heats up. It is then circulated around the elements at the rear which, being hotter than your kitchen, transfer heat into the room. The coolant then goes back into the interior of the fridge. As it does so it expands and cools to below the temperature inside the fridge. Being cooler than the interior, it now absorbs heat from the interior. It then goes through the compressor again and the whole cycle repeats. The net effect is the transfer of energy from inside the fridge to outside – from cold to hot – but at the cost of the energy needed to power the compressor, which is why your fridge doesn't work unless you plug it in.

The energy to power the compressor comes from a power station, which could be a steam engine; a hot furnace in a cold environment. The power station might run off coal or gas, which came from plants, which stored energy from the Sun, which is a hot spot in a cold sky. The stars are the furnaces of the Universe – the ultimate steam engines. At each stage in the flow of energy from the shining stars to the formation of the ice cubes for your (late) afternoon gin and tonic, the overall entropy of the Universe increases as energy flows from hot to cold, and perhaps the cool gin and tonic stimulates 'acts of literary, artistic, and musical creativity that enhance our culture'.

The stars were formed by the gravitational collapse of primordial clouds of hydrogen and helium in the early Universe which, for reasons we do not understand, began in an extraordinarily low entropy configuration. The origin of this special initial state of the Universe – a reservoir of low entropy without which we would not exist – is one of the great mysteries in modern physics.

The concept of entropy proved extremely useful for nineteenth-century steam engine designers because it allowed them to understand that the efficiency of a steam engine depends on the temperature difference between the furnace and the environment. If there is no temperature difference, no net energy can be transferred, and no work can be done. A larger temperature difference allows more work to be done because it allows for more energy to flow without violating the Second Law. But nowhere in the elegant logical edifice constructed by Joule, Clausius and many others, now known as classical thermodynamics, is there any mention of what entropy actually is; it's just a very useful quantity.

What is entropy?

When John Wheeler brought his hot and cold teacups into contact, he was worried about increasing the disorder of the Universe. The link between entropy and disorder was appreciated by James Clerk Maxwell, whose work on electromagnetic theory led Einstein to special relativity. Maxwell realised that the Second Law is different to the other laws of Nature known to the nineteenth-century physicists in that it is inherently statistical. In 1870, he wrote: 'The 2nd law of thermodynamics has the same degree of truth as the statement that if you throw a tumblerful of water into the sea you cannot get the same tumblerful out of the water again.'[28]

In 1877, Ludwig Boltzmann reinforced this idea with a brilliant new insight. Boltzmann understood that entropy puts a number on ignorance; in particular, on our ignorance of the exact configuration of the component parts of a system. Take Maxwell's tumblerful of water. Before throwing it into the sea, we know that all the water molecules are in the tumbler. Afterwards, we have far less idea where they are, and the entropy of the system has increased. This idea is powerfully general – things that shuffle and jiggle will, if left alone, tend to mix and disperse, and our ignorance increases as a result.

Boltzmann's insight connects Clausius's definition of entropy, based on temperature and energy, with the arrangement of the internal constituents of a system. The resulting methodology, which treats matter as being built up out of component parts about which we have limited knowledge, is a branch of physics known as statistical mechanics. The subject is a challenging one, both technically and philosophically. David L. Goodstein begins his textbook *States of Matter* with the following paragraph: 'Ludwig Boltzmann, who spent much of his life studying statistical mechanics, died in 1906, by his own hand. Paul Ehrenfest, carrying on the work, died similarly in 1933. Now it is our turn to study Statistical Mechanics.'[29]

A nice way to see the connection between entropy, temperature and the arrangement of the constituents of a system is to think about a particularly simple physical system – a collection of atoms in a box. The behaviour of atoms is the province of quantum theory, which we'll explore in more detail later. For now, we need only one idea; atoms confined inside a box can only have certain specific energies. We say that the system has a discrete set of 'energy levels'. This is where quantum mechanics gets its name; 'quantised' means 'discrete', as in a discrete set of energies. The lowest possible energy an atom can have is known as the ground state. If all the atoms are in the ground state, the temperature of the box of atoms is zero degrees Kelvin (-273 degrees Celsius). If energy is added, some of the atoms will move to higher energy levels. The parameter that determines how the atoms are distributed among the available energy levels is the temperature. The higher the temperature, the higher up the ladder of available energy levels the atoms can climb (as illustrated in Figure 9.2). The details of how much energy must be transferred into the box to change the configuration of the atoms depends on the types of atom present and the size of the box, but the key point is that there exists a single quantity – the temperature – which tells us how the atoms are most likely to be arranged across the allowed energy levels.

Let's now imagine placing a different box of different atoms in contact with our original box. The details of the energy levels will be different, but what's important is that if the two boxes are at the same temperature, no net energy will be transferred between

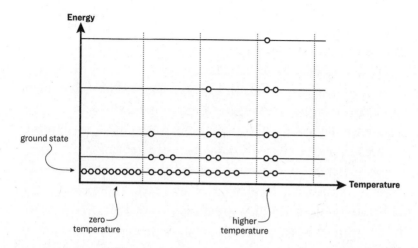

Figure 9.2. Atoms occupying the energy levels for a bunch of atoms in a box. At zero temperature (on the left) all the atoms are in the lowest energy level (the ground state). As the temperature increases (from left to right), atoms increasingly occupy higher energy levels.

the boxes and the internal configurations will not change in a discernible way. This is the meaning of the nineteenth-century concept of temperature. To put it another way, if we place two systems in contact such that energy can be exchanged and nothing happens overall, then the two systems are at the same temperature. This is known as the Zeroth Law of Thermodynamics, because it was an afterthought. The Zeroth Law was always an essential part of the logical structure of classical thermodynamics because it is necessary to pin down the concept of temperature, but it wasn't designated as a law until the early twentieth century, by which time everybody had got so used to speaking of the First and Second Laws of Thermodynamics that they didn't want to change.

Richard Feynman came up with a nice analogy for temperature in his book *The Character of Physical Law*.[30] Imagine sitting on a beach as the clouds sweep in off the ocean and it begins to rain. You grab your towels and rush into a beach hut. The towels are wet, but not as wet as you, and so you can start to dry yourself. You get drier until every towel is as wet as you, at which point there is no way of removing any more water. You could explain this by inventing a quantity called 'ease of removing water' and say that you and the towels all have the same value of that quantity. This doesn't mean that everything contains the same amount of water. A big towel will contain more water than a small towel, but because they all have the same 'ease of removing water', there can be no net transfer of water between them. The reason why an object has a particular 'ease of removing water' is complicated and related to its internal atomic structure, but we don't need to know the details if all we're concerned about is getting dry. The analogy with thermodynamics is that the amount of water is the energy, and the 'ease of removing water' is the temperature. When we say that two objects have the same temperature, we don't mean that they have the same energy. We mean that if we place them in contact their atoms or molecules will jiggle around and collide, just as the molecules in Joule's paddle collided with molecules in the water and imparted energy to them, but if the objects are at the same temperature, the net transfer of energy will be zero and nothing will change on average.

Now recall Wheeler's description of entropy; 'Whatever is composed of the fewest number of units arranged in the most orderly way … has the least entropy.' What does 'order' mean? Imagine we decide to select an atom at random from the box and ask: Which energy level did that atom come from? At zero temperature, we know the answer. The atom came from the ground state. The entropy in this case is zero.* This is what

---

* A technical note. The entropy of a system at zero temperature is zero if the ground state is not degenerate, which means that there are not multiple ground states of the same energy. Solid carbon monoxide and ice are two examples of solids that have degenerate ground states and therefore have a 'residual entropy' at zero temperature because there will still be uncertainty about which state a randomly selected molecule came from.

Wheeler means by 'units' being arranged in an orderly way. We know exactly what we are going to get when we pull an atom out of the box; we are not in the least bit ignorant. If we raise the temperature, the atoms will spread out across the available energy levels, and if we now select an atom at random, we can't be sure which energy level it will come from. The atom could come from the ground state, or from one of the higher energy levels. This means our ignorance has increased as a result of raising the temperature. Equivalently, the entropy is larger, and continues to rise with increasing temperature as the atoms become more distributed among the allowed energy levels.

Temperature, energy and the change in entropy are quantities that appear in classical thermodynamics without any knowledge of the underlying structure of the 'thing' being studied (by 'thing' we mean anything from a box of gas to a galaxy of stars). Thanks to Boltzmann, we now understand that these quantities are intimately related to the constituents of the thing, how those constituents are arranged and how they share the total energy. Temperature, for example, tells us how fast the molecules in a box of gas are moving around on average. Similarly, entropy tells us about the number of possible internal configurations a thing can have. Boltzmann's tombstone bears an inscription of his famous equation for the entropy of a system, which makes the connection with the component parts explicit:

$$S = k_{\mathrm{B}} \log W$$

In this equation, W is the number of possible internal configurations and the entropy, denoted S, is proportional to the logarithm of W. Accordingly, larger W means larger entropy. The logarithm and Boltzmann's constant k_{B} are not important for what follows, other than to note that they allow us to put a precise number on the entropy that also agrees with Clausius's definition in terms of energy and temperature. The important point is that W is the total number of different ways that the component parts of a system could be arranged in a manner consistent with what we know about the system. For atoms in a box, if the temperature is zero, there is only one way the atoms could be arranged, and so W = 1 and the entropy is zero.* If the temperature is raised and some of the atoms hop into higher energy levels, there are more possible arrangements and so W is larger and the entropy is larger.

For a gas in a room, the component parts are atoms or molecules and the things we know about the system might be the volume of the room, the total weight of the gas inside and the temperature. Computing the entropy is then an exercise in counting the different ways the atoms could be arranged inside the room given what we know. One possible arrangement would be that all the atoms, bar one, are sitting still in one corner of the room, while a single lone atom carries almost all the energy. Or maybe the atoms share out the energy equally and are distributed uniformly around the room. And so on. Crucially, there are vastly more ways to arrange the atoms in the room such that the atoms are spread out across the room and share the energy reasonably evenly between them, compared to arrangements where all the atoms are in one corner or the energy is distributed very unevenly. Boltzmann understood that if the energy in the room is allowed to get shuffled around among the atoms because the atoms collide, then all the different arrangements will be more-or-less equally likely. Given that insight and given the numerical dominance of arrangements where the atoms are scattered all over the room, it follows that if we are in a 'typical' room then we are very likely to find the atoms distributed in a roughly uniform fashion. When everything has settled down and things are evenly distributed, we say that the system is in thermodynamic equilibrium. The entropy is then as big as it can be, and every region of the room is at the same temperature.

---

* $\log 1 = 0$. We use '$\log W$' to indicate the natural logarithm of W.

Now, here is why the Second Law embodies the idea of change. If we begin with a system far from equilibrium, which is to say that the component parts are distributed in an unusual way, then as long as the components can interact with each other and share their energy, the system will inexorably head towards equilibrium because that's the most likely thing to happen. We can now appreciate why Maxwell's observation that the Second Law has a statistical element to it was so insightful. The Second Law deals, ultimately, with what is more or less likely, and it is far more likely that a system will head towards thermodynamic equilibrium because there are so many more ways for it to be in thermodynamic equilibrium.

This 'one way' evolution of a system is often called the thermodynamic arrow of time because it draws a sharp distinction between the past and the future: the past is more ordered than the future. In our Universe as a whole, the arrow of time can be traced back to the mysterious highly-ordered, low-entropy state of the Big Bang.

Entropy and information

Suppose we know the precise details of every atom in a room and choose to think in these terms rather than in terms of the volume, weight of gas and temperature. Then the entropy would be zero because we know the configuration exactly. This means that an omniscient being has no need for entropy. For mortal beings and physicists, however, the vast numbers of atoms in rooms and other large objects makes keeping track of their individual motions impossible and, as a result, entropy is a very useful concept. Entropy is telling us about the amount of information that is hidden from us when we describe a system in terms of just a few numbers. Seen in this way, entropy is a measure of our lack of knowledge, our ignorance. The connection between entropy and information was made explicit in 1948 by Claude Shannon in one of the foundational works in what is now known as information theory, which is central to modern computing and communications technology.

Returning to our gas-filled room, there will be many possible configurations of the atoms inside that are consistent with our measurements of the volume, weight and temperature. The logarithm of that number, according to Boltzmann, is the entropy. Importantly, though, the atoms are actually in a particular configuration at some moment of time. We just don't know what it is. Let's imagine that we make a measurement and determine precisely which configuration. What have we learnt? To be more specific, exactly how much information have we gained? Following Shannon, the amount of information gained is defined to be the minimum number of binary digits (bits) required to distinguish the measured configuration from all the other possible configurations. Imagine, for example, that there are only four possible configurations. In binary code, we would label those configurations as 00, 01, 10 and 11. That means we gain two bits of information when we measure it. If there are eight possible configurations, we would label them 000, 001, 010, 011, 100, 101, 110 and 111. That's three bits. And so on. If there were a million possible configurations, it would take us some time to write all the combinations out by hand, but we don't need to because there is a simple formula that tells us how many bits we'd need. If the number of configurations is W, the number of bits is:

$$N = \log_2 W$$

This is very similar to Boltzmann's formula for the entropy of the box of gas. If you know a little mathematics, you'll notice that the logarithm is now in base 2 rather than the natural logarithm in Boltzmann's formula, but that just leads to an overall numerical factor.* The key point is that the information gained in our meas-

---

* log 2 = 1.4427.

urement of the precise state of the gas is directly proportional to the entropy of the gas before the measurement. Specifically:

$$N = \frac{S}{1.4427 k_{\text{B}}}$$

This is the key to understanding the fundamental importance of entropy. Entropy tells us about the internal structure of a thing – it is intimately related to the amount of information that the thing can store and as such it is intimately linked to the fundamental building blocks of the world. It is a window into the underlying structure of cups of tea, steam engines and stars. And, if we follow Bekenstein and associate an entropy with the area of the event horizon of a black hole, it is a window into the underlying structure of space and time.

The entropy of a black hole

The immediate issue is that black holes as we have described them so far have no component parts. They are pure spacetime geometry and quite featureless. Superficially, therefore, a black hole would appear to have zero entropy. Throw a couple of cups of tea into a black hole and its mass will increase, but that's all and so the entropy should still be zero. This was Wheeler's point. To save the Second Law, in true Eddingtonian spirit, Bekenstein guessed that black holes must have an entropy, and that must be proportional to the area of the horizon. Bekenstein did more than just guess though. In an ingenious back of the envelope calculation, he also estimated the numerical value of the entropy of a black hole and discovered something very deep.

Let's imagine that we want to drop a single bit of information into a black hole. How can we think of achieving this? One answer is to drop a single photon into the black hole. A photon is a massless particle of light, and one photon can store one bit of information. We can think of a photon as spinning clockwise (0) or anticlockwise (1), which means it can represent one bit. Each photon also carries a fixed amount of energy proportional to its wavelength. This relationship between energy and wavelength was first proposed by Einstein in 1905 and is a key feature of quantum theory. Long-wavelength photons have smaller energy and short-wavelength photons have higher energy. This is the reason UV light from the Sun can be dangerous but the light from a candle can't: UV photons have short wavelengths and carry enough energy to damage your cells, whereas candlelight photons have longer wavelengths and don't carry enough energy to cause damage. As a general rule, the position of a photon cannot be resolved to distances smaller than its wavelength. This means we would like to drop a photon with a wavelength roughly equal to or smaller than the Schwarzschild radius into the black hole, since longer wavelength photons would typically reside outside the hole. Now we can calculate the largest possible number of such photons we can fit inside the black hole. This should give us a crude estimate of the maximum number of bits it can store, which is the entropy.* We go through the calculation in Box 9.1. The answer for the number of bits hidden in a Schwarzschild black hole is:

$$N = \frac{c^3}{8\pi Gh} A$$

where A is the area of the event horizon.† One very interesting thing about this equation is the collection of numbers in front of the horizon area, A. This combination of the speed of light, c, Newton's gravitational constant, G, and Planck's constant, h, a

---

* We explain why a black hole has the largest possible entropy shortly.

† A more careful calculation, which properly accounts for the quantum physics, gives an entropy equal to the horizon area divided by (4 x Planck length squared), which differs from our estimate by a numerical factor.

number that lies at the heart of quantum theory, is well known to physicists. It is the square of the so-called Planck length. We've described the significance of the Planck length in more detail in Box 9.2. The short version is that it is the fundamental length scale in our Universe and the smallest distance we can speak of as a distance. The suggestion is that the entropy of a black hole, which is to say the number of bits of information it hides, can be found by tiling the event horizon in Planck-length-sized pixels and assuming that the black hole stores one bit per pixel. This is illustrated in Figure 9.3.

It's hard to overstate what an intriguing result this is. What is the nature of these Planckian pixels and why are they tiling the horizon when, according to general relativity, the horizon is just empty space? Recall, an astronaut freely falling through the horizon should experience nothing out of the ordinary according to the Equivalence Principle. And yet Bekenstein's result suggests

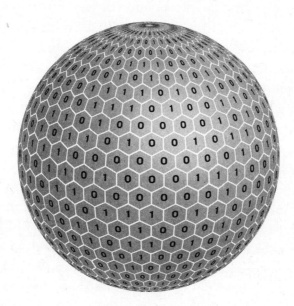

Figure 9.3. The horizon of a black hole, with an imaginary tiling of Planck-area-sized cells. Remarkably, the total number of cells is equal to the entropy of the black hole.

that they encounter a dense collection of bits. Furthermore, why should the information capacity of a black hole be proportional to the area of its horizon rather than its volume? How much information can be stored in a library? Surely the answer must depend on the number of books that fit inside. For a black hole library, however, it seems as if we are only allowed to paper the outside walls with the pages of the books. It is as if the interior doesn't exist.

One might wonder whether the black hole might be missing a trick from an information storage perspective, but a simple argument suggests that a black hole of a given mass has the maximum possible information storage capacity (entropy) for that mass. Imagine dropping an object into a Schwarzschild black hole. To obey the Second Law, the black hole entropy must increase by at least the entropy of the object it swallows. The area of its event horizon will grow accordingly, but the area increase depends only on the mass of the object because the area of the horizon is proportional only to its mass. Now imagine instead dropping a super-high entropy object with the same mass as before into the black hole. The horizon area will increase by precisely the same amount, proportional only to the mass of the object. This means that as we add mass to the black hole, it must increase its entropy by the largest possible amount. It is as if objects thrown in get completely scrambled up, to guarantee that our ignorance is maximised.

A black hole therefore has the largest possible entropy. It can store the maximum possible amount of information in a given region of space, and the amount of information measured in bits is given by the surface area of the region in Planck units. This hints at something deeply hidden; everything that exists in a volume of space can be completely described by information on a surface surrounding the region. This is our first encounter with the holographic principle.

BOX 9.1. Black hole entropy

Roughly speaking, only photons whose wavelengths (as measured by a distant observer) are less than the Schwarzschild radius can fit inside the hole. According to quantum physics, a photon has an energy $E = hc/l$, where l is the wavelength and h is Planck's constant. Therefore, the smallest possible photon energy is $E = hc/R$, where R is the Schwarzschild radius. The total energy of a black hole of mass M is, according to the famous Einstein relation, Mc^2. The maximum number of photons we can fit inside the hole is:

$$N = \frac{Mc^3}{hc/R} = \frac{Mc^2R}{h}$$

Now, the Schwarzschild radius $R = 2GM/c^2$, which means we can write:

$$N = \frac{R^2c^3}{2Gh}$$

The horizon area $A = 4\pi R^2$, and therefore:

$$N = \frac{c^3}{8\pi Gh} A$$

You may wonder if more information can be stored using other types of particle (electrons also spin and can be used to encode bits). Unlike photons, other particles carry mass which means fewer of them can fit inside the hole.

BOX 9.2. The Planck length

The Planck length is a combination of three fundamental physical constants; Planck's constant, Newton's gravitational constant and the speed of light. The Planck length is a very tiny length. The diameter of a proton is 100,000,000,000,000,000,000 Planck lengths. Max Planck first introduced his eponymous unit in 1899 as a system of measurement that depends only on fundamental physical constants. This is preferable to using things like metres or seconds, which reflect the vagaries of history and have more to do with the size of humans and the orbit of our planet than the underlying laws of Nature. The strength of gravity, the behaviour of atoms and the universal speed limit, however, are independent of humans. If we encountered an alien civilisation and asked them to tell us the area of the horizon of the M87 black hole in Planck units, they would come up with the same number that we do. In a formula, the Planck length is given by:

$$l_\mathrm{p} = \sqrt{\frac{hG}{2\pi c^3}} \approx 10^{-35} \text{ metres}$$

It is believed to be the smallest distance that makes any sense: smaller than this, it is likely that the idea of a continuous space breaks down.

History repeats itself

There is a parallel between Bekenstein's proposal for the entropy of a black hole and the development of nineteenth-century statistical mechanics. When Boltzmann died in 1906, aged 62 years, his atomistic explanation of the Second Law was still not universally accepted. Led by the greatly influential German physicist

Ernst Mach, many scientists still doubted the very existence of atoms. Mach's objections were initially of a philosophical nature, but they developed a momentum in part because Boltzmann's work led to a good deal of confusion that Boltzmann himself struggled to dispel. The argument became centred around the statistical nature of the Second Law. According to Boltzmann, if matter is made of atoms all moving around then it is overwhelmingly likely that entropy will increase, but it is not guaranteed. For example, there is a near-vanishingly small probability that all the atoms in a room will end up clustered in one corner. Mach and his followers felt that a fundamental law of Nature should not be statistical. 'Entropy almost always increases' didn't sound authoritative enough, especially since Clausius's formulation of the Second Law has no 'almost' about it. Today, we know Boltzmann was right – the Second Law does involve an element of probability.

A similar debate now centres around the physical significance of the thermodynamic behaviour of black holes. If we accept the idea that the entropy of a black hole is signalling the presence of 'moving parts' of some sort, the astonishing implication is that general relativity is underpinned by a statistical theory just as classical thermodynamics is underpinned by statistical mechanics. This means that we should regard spacetime as an approximation; an averaged-out description of the world akin to the description of a box of gas in terms of temperature, volume and weight. In 1902, the pioneer of statistical mechanics Josiah Willard Gibbs wrote: 'The laws of thermodynamics ... express the approximate and probable behaviour of systems of a great number of particles, or, more precisely, they express the laws of mechanics for such systems as they appear to beings who have not the fineness of perception to enable them to appreciate quantities of the order of magnitude of those which relate to single particles ...' Could it be that, in the first decades of the twenty-first century, we find ourselves in a similar position, as beings who have not the fineness of perception to appreciate the underlying structure of space and time?

When Bekenstein made his suggestion that black holes have an entropy, however, there was one huge fly in the ointment. As we've seen, entropy and temperature go hand in hand, and the assignment of a thermodynamic entropy to a black hole requires it to have a temperature. But for an object to have a temperature it must be able to emit things as well as to absorb them. Temperature, after all, can be defined in terms of the net transfer of energy between objects – in Feynman's analogy, the 'ease of removing water'. If it is not possible to extract anything from a black hole then the temperature must be zero. And, as everyone knew in 1972 and as general relativity makes abundantly clear, nothing can escape from a black hole.

Then, in 1974, everything changed. Along came Stephen Hawking with a short paper entitled 'Black Hole Explosions?'

10

HAWKING RADIATION

'Bardeen, Carter and I considered that the thermodynamical similarity was only an analogy. The present result seems to indicate, however, that there is more to it than this.'

Stephen Hawking[31]

Stephen Hawking's paper triggered a revolution in theoretical physics that is still ongoing. He discovered that quantum theory predicts that a black hole will emit radiation as if it were an ordinary object with a temperature. The immediate suggestion is that the law of gravity should be regarded as a statistical law and that quantum effects lead to an elemental randomness in the geometry of space. Today, we do not know what that randomness corresponds to. It remains the holy grail of theoretical physics. But we have travelled a long way since 1974. The remainder of the book is about the quest to understand the deep origins of black hole thermodynamics; a quest that is edging us ever closer to a new theory of space and time.

The laws of black hole mechanics

In 1973, Bardeen, Carter and Hawking published a paper entitled 'The Four Laws of Black Hole Mechanics' in which they drew the following analogy between the laws of classical thermodynamics and the properties of black holes:[32]

	Thermodynamics	**Black hole mechanics**
0th law	Temperature (T) is constant	Surface gravity (k) is constant
1st law	$dE = T\,dS$	$dE = \dfrac{k}{2\pi}\dfrac{dA}{4}$
2nd law	Entropy (S) increases	Horizon area (A) increases
3rd law	T cannot fall to zero	k cannot fall to zero

Even if you don't understand the symbols, the similarity is striking. One set of laws can be obtained from the other by swapping 'temperature (T)' with 'surface gravity (k)' (divided by 2π), and 'entropy (S)' with 'area (A)' (divided by 4).

Let's start with the Zeroth Law, which as we saw in the last chapter formally anchors the concept of temperature. A system such as a box of gas is in equilibrium if everything has settled down and nothing is happening. This means that all parts of the system have the same temperature. For a black hole, the corresponding quantity is the surface gravity, k. This has the same value everywhere on the event horizon of a black hole that has settled down after swallowing a planet, for example. The surface gravity tells us how difficult it is to resist the pull of gravity just above the event horizon. Imagine that, in an admittedly surreal turn of events, an astronaut decides to conduct a spacewalk, carrying a fishing rod, close to a black hole. On the end of the fishing line is a trout of mass M. The astronaut lowers the trout down until it is dangling just above the horizon and measures the tension on the fishing line. The tension will be kM, where k is the surface gravity of the black hole.* For a perfectly spherical (Schwarzschild) black hole it is perhaps obvious that the surface gravity should not vary

---

*   The surface gravity is inversely proportional to the mass of the black hole.

as one moves around the horizon. For a rotating (Kerr) black hole this is not obvious at all. A proof is provided in the Bardeen, Carter and Hawking paper.

The First Law expresses the conservation of energy. It says that if we add energy (dE) into a system at given temperature (T) we increase the entropy (dS). The corresponding law of black hole thermodynamics states that if we drop an amount of energy (dE) into a black hole that has a surface gravity (k) then the surface area of the horizon will increase (dA). If we are tempted to identify the surface gravity with the temperature, then we might also be tempted to identify the surface area with the entropy as Jacob Bekenstein proposed. That temptation is made all the greater by the Second Law which, for a black hole, is a statement of Hawking's discovery that the area of the event horizon always increases. This apparent link between the purely geometric concept of area and the information content of a system is very unexpected.

The Third Law is interesting too. In classical thermodynamics, it states that it is not possible to cool something down to zero temperature in a finite series of steps. One way to see this is to think about a fridge again. As the temperature inside the fridge gets closer and closer to zero, the efficiency of the fridge also gets closer and closer to zero. That's because removing energy from something at very low temperature involves a huge entropy change, which has to be accounted for by sending an appropriately huge amount of energy out into the environment. Ultimately, the fridge would have to do an infinite amount of work to transfer the last drips of energy from inside to outside and cool the interior to absolute zero. For a Schwarzschild black hole, we could make the surface gravity go to zero by making its mass infinite, which obviously requires an infinite amount of energy. For a Kerr black hole, the situation is different. It is possible to reduce the surface gravity by throwing matter into the hole, if the matter is rotating. At first sight, it looks as if this could be used to dodge the Third Law and get the surface gravity down to zero, but that isn't the case. Remarkably, it turns out that as the surface gravity gets smaller, it becomes harder to throw stuff into the hole. The matter either misses the hole or gets repelled by it.

Bardeen, Carter and Hawking conclude their 1973 discussion with the following comments: 'It can be seen that k is analogous to temperature in the same way that A is analogous to entropy. It should be emphasised, however, that k and A are distinct from the temperature and entropy of the black hole. In fact, the effective temperature of a black hole is absolute zero.'

A few months later in his 1974 paper Hawking disagrees with himself, which is one of a scientist's most important abilities: '… it seems that any black hole will create and emit particles such as neutrinos or photons at just the rate one would expect if the black hole was a body with temperature …' His more detailed 1975 follow-up, 'Particle Creation by Black Holes', contains the following equation for the temperature of a black hole:[33]

$$T = \frac{k}{2\pi}$$

which states that black holes have a temperature equal to their surface gravity divided by 2π. Hawking has now realised that the similarity between the laws of thermodynamics and the laws of black hole mechanics is not just an analogy. Rather, it appears to be an exact correspondence and black holes are thermodynamic objects. As Hawking writes, 'if one accepts that black holes do emit particles at a steady rate, the identification of $k/2\pi$ with the temperature and $A/4$ with the entropy is established and a Generalised Second Law confirmed.'

Hawking's discovery that black holes emit particles is of profound importance, not least because it suggests that the origin of the law of gravity is statistical. This is the shock wave that reverberated through the theoretical physics community in the early

1970s. Just as the concepts of temperature and entropy for a box of gas emerge from a hidden microscopic world composed of lots of little things jiggling and reconfiguring, so it seems do the laws of gravitation. But how is it possible for a thing from which nothing should escape to glow like a hot coal? To understand Hawking's discovery we need to turn to quantum theory, and to the physics of nothing.

Hawking radiation

We haven't met quantum theory in any detail yet because we've been discussing general relativity, which is a classical theory. Classical theories describe a reality that fits nicely with our intuitive mental picture of the world. The Universe is composed of particles, fields, and forces. At any moment there is a single configuration of the Universe, and this evolves in a predictable way into a new configuration as things interact with each other in the arena of spacetime. General relativity tells us how spacetime reacts to the particles and fields and how the particles and fields react to spacetime.

Quantum mechanics is different. It describes a world of probabilities and of multiple possibilities. For example, when a particle moves from A to B, quantum mechanics says we must take all possible paths into account if we are to make predictions that agree with experimental observations. In classical physics, the particle follows a single path, but this is not so in quantum physics.

A central difference between classical theory and quantum theory is the unavoidable appearance of probabilities in the description of Nature. This is encapsulated by the famous Heisenberg Uncertainty Principle, which states that we cannot simultaneously know the precise position and momentum of a particle. If we know with high precision where a particle is, we know with less precision how fast it is moving. The consequence is that we cannot predict with certainty where a particle will be in the future, even if we know everything it is possible to know about its current state. Rather the theory gives us a list of probabilities for possible future locations. This is not because of a lack of knowledge or skill on our part. It's the way Nature is. Very importantly, though, we can still predict how the so-called quantum state of the particle changes over time. Precise knowledge of the quantum state provides us with a list of probabilities to find the particle in some region of space, and we can predict with precision how this list of probabilities changes over time, though we can never say for sure where the particle will be. We are therefore not able to know precisely where an electron will be at some moment, or precisely how much energy the electromagnetic field will carry in some region of space. We can only know the probability that a particle will be somewhere, or the probability that a field will be in a certain configuration. It is this inherent uncertainty in the configuration of particles and fields that ultimately leads to Hawking radiation.

We should emphasise that, as far as we can tell, all of Nature is quantum mechanical. Quantum theory is as venerable as general relativity and underpins not only our understanding of atoms and molecules and all of chemistry and nuclear physics, but also modern-day electronics. For example, the semiconductor transistor used in their billions in modern electronic devices is an inherently quantum mechanical device. We live in a quantum universe.

The quantum vacuum

There are certain words in colloquial usage that mean something very different in physics. Vacuum is one such word. The important feature of our quantum universe that leads to Hawking radiation is that the vacuum of empty space is not empty. It's natural to picture a vacuum as being empty – devoid of all particles and fields – but this is not correct. The vacuum can't be empty, because 'empty' is a precise statement about the energy and configuration of the fields, and quantum theory does not allow that. The vacuum is therefore an active place with a complex structure. There is no way to isolate a region of space and suck all the particles out of it to leave it perfectly empty. Roughly speaking, the vacuum is to humans as water is to a fish: it is an ever-present backdrop to our everyday experience. Particles can be thought of as excitations of the vacuum – ripples in the vacuum sea – and quantum theory describes a sea that always ripples.

One way to picture the quantum vacuum is to imagine particles constantly popping into it and disappearing again a fleeting instant later. These ghosts – the momentary flickers of particles – are known as vacuum fluctuations or 'virtual' particles. If we could somehow freeze time and peer with high-resolution vision deep into any region of space, we'd see these particles, not as fleeting ghosts but as real particles. But that's what happens to time and space close to the horizon of a black hole, as viewed from the outside. The black hole behaves like a magnifying glass, freezing

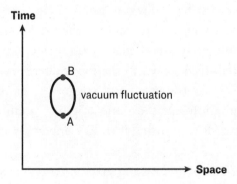

Figure 10.1. A vacuum fluctuation. A pair of particles emerges (at A) from the vacuum for a fleeting instant of time before recombining at B. One of the particles can be pictured as having negative energy such that the sum-total of the two particle energies is zero.

time and changing the way we view the vacuum fluctuations. Virtual particles from one point of view can be as real as the particles that make up our bodies from another.

Virtual particles emerge from the vacuum in pairs, and if you could watch them flickering in and out of existence in front of your nose you'd observe that one of the ghostly particles would have positive energy and the other negative energy. In the normal scheme of things, the particles come back together again in a very short time and the energy is repaid, so that the total energy of the vacuum remains unchanged on the average. This process, however fantastical it may sound, is familiar. When you switch on a fluorescent light, atoms of the vapour inside the tube are supplied with energy and jump into an excited state. This means that electrons inside the atoms now occupy energy levels above the ground state. We met these energy levels in our discussion of temperature and entropy in the previous chapter: one can think of a single atom rather like a box, containing electrons that are distributed among the available energy levels. The electrons occupy the higher energy levels for a while before dropping down to lower levels. In doing so, they emit photons which carry the energy away and cause the fluorescent tube to glow. For a time, the reason for the electrons dropping back down to lower energy levels in the atom was not understood, and physicists called it 'spontaneous emission'. It is now understood that vacuum fluctuations cause the electrons to fall back down to the lower energy levels inside the atom. They 'tickle' the atom and trigger the emission of light. Hawking radiation has the same origin. Vacuum fluctuations tickle the black hole, causing it to lose energy by emitting particles.

In his 1975 paper, Hawking gives a heuristic explanation for the origin of black hole radiation. As he is careful to note, these physical pictures are not meant to be a rigorous argument and 'the real justification of the thermal emission is the mathematical derivation'. Nevertheless, as Hawking appreciated, pictures are very useful for developing understanding. Here's Hawking's picture.

Let's station ourselves outside the event horizon of a black hole and focus on the vacuum fluctuations close to the horizon. From this perspective, the vacuum fluctuations can be disrupted such that one of the particles escapes recombination with its partner. The reason is that the negative energy particle in the fluctuating pair can be inside the horizon, where it can exist until it reaches the singularity. The possibility of negative energy particles existing inside a black hole is something we encountered in the Penrose process for extracting energy from a black hole. In that case, the fact that space and time 'switch roles' inside the ergosphere was responsible. That same reversal of the roles of space and time is why the negative energy particle inside the horizon will reduce the mass of the black hole while its partner can head outwards into the Universe and appear as Hawking radiation.*

For a Schwarzschild black hole, it is possible to express the equation for the temperature of a black hole in terms of the mass M of the black hole, rather than the surface gravity. The result is that†

$$T = \frac{\hbar c^3}{8\pi G M k_B}$$

This wonderful equation reveals the marriage of quantum theory and general relativity that is present in Hawking's calculation, and it establishes that the laws of black hole mechanics *are* the fundamental laws of thermodynamics in disguise. This is what convinced physicists that it is correct to treat black holes as thermodynamic objects that can store information and exchange

---

* Not all the positive energy particles will travel away from the hole. Some will fall into it and end up in the singularity. The point is that some of the particles can escape.

† $\hbar$ is Planck's constant, h, divided by 2π.

Figure 10.2. The temperature of a Schwarzschild black hole, written
on Stephen Hawking's memorial stone in Westminster Abbey.
Also on plate 13.

energy with the Universe beyond the horizon. Stephen Hawking's
calculation of the temperature of a black hole is so important to
our understanding of the Universe that it is now literally written
in stone on the floor of Westminster Abbey.

11

SPAGHETTIFIED AND VAPORISED

'In microphysics, however, the information does not sit out there. Instead, Nature in the small confronts us with a revolutionary pistol, "No question, no answer." Complementarity rules.'

John Archibald Wheeler[34]

The 'black holes are a bit like atoms in the way they emit radiation as a consequence of being tickled by the vacuum' picture is a good one, but it disguises a major difference between the emission of light from everyday hot objects and the emission of Hawking radiation. The difference can be traced to the fact that it is gravitational effects that make the vacuum fluctuations real. This unique production mechanism gives rise to three properties of Hawking radiation that, taken together, appear nothing short of bamboozling.

1. Someone falling freely near to the horizon of a large black hole will not encounter any radiation.
2. Someone accelerating, so that they hover just above the horizon of a large black hole, will be vaporised by a flux of very hot radiation.
3. Someone far away from the black hole will experience a flux of cool radiation which appears to have been emitted by a glowing object at the Hawking temperature.

Let's deal with each of these in turn.

1. That someone falling freely near the horizon of the large*
   black hole should not encounter any radiation is not
   hard to understand. This is Einstein's Equivalence
   Principle. A freely falling observer feels as if they are at
   rest in plain old flat spacetime. From their perspective
   therefore, they experience the vacuum fluctuations (those
   particle–antiparticle pairs) in the same benign way as
   someone floating far away from the black hole. As a
   result, they float onwards unawares until, inexorably,
   they are spaghettified as they approach the singularity.

2. In contrast, someone hovering just above the horizon will
   encounter the positive energy part of the vacuum
   fluctuations. From their perspective they will be
   bombarded by a flux of real particles, separated from their
   partners by the geometry of spacetime. A very similar
   effect occurs even in flat spacetime, as was appreciated by
   Paul Davies and Bill Unruh in the mid-1970s.†[35] In the
   Davies–Unruh effect, an accelerating rocket ship far from
   any gravitating objects will experience a thermal bath of
   particles and the temperature of the bath is proportional
   to the acceleration. The Equivalence Principle can be
   invoked to say that the same thing will happen to a
   rocket that accelerates to maintain a fixed position close
   to the horizon of a black hole. In that case, because
   spacetime in the immediate vicinity of the rocket is
   approximately flat, the rocket's experience is just like that
   of a rocket accelerating in flat spacetime. And therefore,
   as a result of being immersed in a hot bath of particles,

---

* Choosing a large black hole means that tidal effects are small at the horizon.
† Often called the Fulling–Davies–Unruh effect to acknowledge the earlier work in
  1973 by Stephen Fulling.

the rocket heats up. If it is close enough to the horizon, it will be vaporised.

3. A freely falling observer far from the hole will detect Hawking radiation, but the particles they receive will be benign, with energies characteristic of an object radiating at the Hawking temperature. This is as expected – it is what Hawking predicted. Another way to understand how Hawking radiation arises from this far away point of view is to think about the tidal gravity of the black hole. Tides on Earth are raised because the gravitational pull of the Moon varies across the Earth. The result is that the ocean surface, and to a very small extent the Earth itself, is distorted by the varying gravitational field. On Earth, tidal gravitational effects are only felt over very long distances – at widely separated points on the Earth's surface – because they are caused by the *difference* in the gravitational pull of the Moon in two places. There is no measurable difference in gravitational pull over a couple of metres, which is why the Moon does not raise tides in your bath. Hawking radiation arises because the gravitational pull of the black hole varies across a vacuum fluctuation. For this to be a large enough effect to make the particles real, the vacuum fluctuation must be separated by approximately the size of the black hole. From the distant vantage point, an observer therefore sees a flux of long-wavelength (low-energy) particles.

Each of these three experiences are legitimate descriptions from different viewpoints, yet at first sight they appear to be mutually contradictory. Let us make the point even more starkly. Suppose you jump into a large black hole. From your perspective you are doomed to spaghettification as you approach the singularity, but you will cross the event horizon with no drama. Your friend in a rocket ship outside of the hole will never see you cross the horizon, but they will see you get closer and closer. That much we know from general relativity. They might decide to lower a thermometer down to measure the temperature at your position close to the horizon. The thermometer, hovering above the horizon, will experience the vacuum fluctuations as in case 2 above and will therefore be immersed in a hot bath of real radiation. From this vantage point, the near-horizon region is a scorching hot place. Your friend may conclude that you got burnt to a crisp outside the black hole and never made it through the horizon.

It seems there is little way out here. Should we abandon the Equivalence Principle, the very foundation of general relativity, and conclude that the horizon is a hot and dangerous place? Or should we insist that a freely falling observer must cross the horizon with no drama and conclude that there is something wrong with our quantum physics analysis of the vacuum? If we take one or other of these positions, we are required to ditch a core element of either general relativity or quantum theory.

There is a third way. A beautiful expression of centrism. It is possible that both perspectives are correct. From the outside perspective, the black hole has a scorching hot, impenetrable atmosphere which vaporises everything that approaches. Yet, according to in-fallers, the horizon is a complete non-entity and they pass through unharmed into the interior. A person can be both spaghettified *and* vaporised: Spaghettified from their own perspective and vaporised according to outsiders. This idea is known as 'black hole complementarity'.[36]

From the outside, nothing is ever observed to fall into a black hole. We might say that the interior lies beyond the end of time for someone lurking outside. Stuff just falls into a hot atmosphere where it gets burnt up. From the outsiders' point of view, it seems that a black hole is not so different to a hot, glowing coal.

According to the black hole complementarity paradigm, none of this is contradicted by the narrative of someone who falls into the hole, although the story they tell is different. They get to explore the interior and meet with the singularity. Crucially, however, they are not able to communicate any of this to the outsiders once they have crossed the horizon. Equally crucially, the outsiders are unable to inform the in-faller that they have been burnt up. A contradiction is avoided because the outsiders and in-fallers never get together to compare notes. This sounds like a proper bodge, to use our native vernacular, but it is a serious proposition.

You might immediately raise the following objection. What if the outsider collects some of the ashes of the in-faller and then jumps in after their friend to show them the evidence that they were incinerated? Being confronted with one's own ashes would be a disconcerting experience, even for the most ardent advocate of black hole complementarity. The way out of this logical impossibility, according to complementarity, is that it takes time for the outsider to collect the evidence that the in-faller has burnt up. By the time the outsider has gathered the evidence and jumped in across the horizon, their friend has already been spaghettified and cannot therefore be presented with the evidence of their own demise. However bizarre it may seem, the picture we've just outlined appears to be essentially correct. The route to this realisation can be traced back to the 1980s and a very simple question.

The information paradox

The title of Stephen Hawking's initial paper on Hawking radiation was 'Black Hole Explosions?' The reason for this eye-catching title is that the temperature of the black hole is predicted to rise as it shrinks, causing it to radiate more fiercely and shrink faster until it completely disappears in a flash of radiation.* 'Faster' is perhaps the wrong word to use, introducing an unwarranted sense

---

* You can see this directly from Hawking's formula – the temperature of a black hole is inversely proportional to its mass.

of urgency to the process. You can pop a few numbers into Hawking's formula to calculate the temperature of a typical stellar mass black hole – let's say around five times the mass of the Sun. Doing so reveals that the temperature is ten billionths of a degree above absolute zero. That's very much colder than the Universe today, which has cooled down to around 2.7 degrees above absolute zero in the 13.8 billion years since the Big Bang. At this moment in cosmic time, black holes are more like Wheeler's iced teacups and, in accord with the Second Law of Thermodynamics, they are absorbing energy as they float in the relatively hot bath of the cosmos. There will come a time, however, as the Universe continues to expand and cool, when they become glowing hot spots in the ever-chilling sky and then they will begin to evaporate. A typical solar mass black hole will have a lifetime of approximately 10^{69} years, which is a very long time. At first sight, therefore, it might appear that we don't need to worry about black hole explosions; their lifetimes are surely as close to infinite as makes no odds. If we stopped here, however, we would miss what is quite possibly the most important revolution in theoretical physics since Einstein's time. The revolution was triggered by the following question: 'Do black holes destroy information?'

Imagine that a book falls into a black hole. Over incomprehensible time scales, the black hole will gradually evaporate away as it emits Hawking radiation until it vanishes in a final burst of radiation. All that remains will be the Hawking radiation. Crucially, Hawking's calculation makes a definite prediction about the nature of this radiation: it is thermal, which means that the radiation encodes no information at all. In other words, when the black hole has vanished, it is as if the book never existed. The information it contained has been erased from the Universe. In fact, all the information about everything that ever fell into the black hole, including the details of the collapsing star out of which it originally formed, will also be erased. Instead, all that remains is a bath of featureless, thermal radiation.

So what? In his book *The Black Hole War*, Leonard Susskind describes the moment in a small seminar room in an attic in San Francisco in 1983 when Hawking first made the claim that information is destroyed by black holes.[37] The soon-to-be Nobel Prize winner Gerard 't Hooft 'stood glaring' at the blackboard for an hour after Hawking's talk. 'I can still see the intense frown on Gerard's face and the amused smile on Stephen's.' 't Hooft was glaring because the laws of physics as we currently understand them preserve information. If we know the precise state of something at a given moment in time then, in principle, we can predict precisely what it will do in the future and know what it was doing in the past. This is determinism, the fundamental idea that the Universe evolves in a predictable way. All the known laws of physics deliver deterministic evolution. They take a system, be it a box of gas or star or galaxy, and evolve it uniquely into a single, well-defined configuration at some time in the future. And because things evolve in a unique and predictable way, the laws also allow us to calculate precisely what the system was like at any time in the past.* But what of a Universe that contains black holes? Galaxies contain supermassive black holes at their cores, and those black holes swallow things. If the black hole subsequently disappears in a puff of information-free Hawking radiation, then it would be impossible in the far future to reconstruct any details about anything that had fallen in. It would in fact be impossible to deduce that the black hole ever existed at all, because it would have erased all trace of itself. This problem has become known as the black hole information paradox.

Figure 11.1 shows a Penrose diagram that illustrates the problem. It is obtained by sewing together two spacetime geometries: the Schwarzschild spacetime corresponding to the period that the black hole exists (Figure 8.3) and flat (Minkowski) spacetime

---

* This is also true for the quantum evolution of the state of a system, even though the outcomes of individual experiments are not determined.

when the black hole has gone. The singularity disappears with the black hole and the wavy line representing the end of time therefore ends before future timelike infinity at the upper tip of the diagram. We have no idea what happens at the rightmost point of the singularity, since this is the event where the black hole disappears and quantum gravity effects are important. But as we'll see, and contrary to the expectations of many experts in the field until very recently, it turns out that we don't need to know to solve the information paradox.

The shaded region of the diagram corresponds to times after the evaporation of the black hole, where spacetime is flat and there appears to be no trace of anything that crossed the event horizon. To see this, draw a light ray (a 45-degree line) from any point inside the event horizon and you'll see that it ends on the singularity. The region behind the horizon remains a prison from which there is no escape because it is causally disconnected from the Universe outside; there is no spacetime directly above the singularity in Figure 11.1. The only thing that survives into the future when the black hole has disappeared is Hawking radiation. We've drawn a wiggly arrow to indicate the worldline of the last Hawking particle to be emitted as the black hole vanishes – it heads off happily to future lightlike infinity.

According to Hawking, the radiation is featureless and the black hole therefore erases all trace of everything that fell into it. Where did everything go? If the hole did not evaporate, we could at least offer the vague statement that 'it fell into the singularity and we don't really understand that place'. But after the hole evaporates, there isn't any singularity anymore – there is no hiding place.

Black hole complementarity offers an apparently simple resolution to the paradox, at least from the point of view of an outside observer. Since nothing is ever seen to fall through the horizon, nothing is ever lost. Let's think again about the book we tossed into the black hole, we have drawn its worldline in blue on the

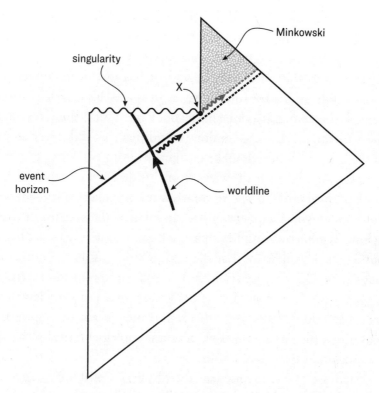

Figure 11.1. Penrose diagram for an evaporating black hole. The
singularity disappears after the last Hawking particle (the grey wiggly
arrow [orange on plate 14]) is emitted (at point X). The slightly curved
line is the worldline of a book thrown into the hole and the darker
wiggly arrow (red on plate 14) is another Hawking particle. Both
Hawking particles follow their respective dotted worldlines and end up
at future lightlike infinity. Also on plate 14.

figure. From the perspective of the outside observer, according to
complementarity, the book is incinerated on the horizon and its
ashes are returned to the Universe as Hawking radiation (as illustrated by the red wiggly arrow). This is no different to burning a
book on a campfire. If we burn a book, the information contained
within can in principle be recovered if we make precise enough
measurements of all the ashes and gases and embers that emerge.
In practice, this is not possible, but practicality is not a word that

concerns theoretical physicists. The point is that it is possible in principle. The information contained in the book got scrambled up during the burning process, but it wasn't destroyed. We would claim that nothing has disappeared, it has just been converted from words on a page into particles in space. It's clear from the Penrose diagram that as long as the book burns up before it crosses the horizon, it is possible to draw worldlines that link every atom in the book to future timelike or lightlike infinity.

Contrast this with the view from inside the black hole. The book is spaghettified as it approaches the singularity. Though we don't know what happens at the singularity, the fact that it lies behind the horizon implies that the book is not able to get back out again. It is destroyed at the end of time inside the horizon. But that is OK because from the external viewpoint information is preserved. The stories in the book are written in the Hawking radiation and will always be there, in principle, for the super-beings of the future to read. If we accept complementarity, both the internal and external viewpoints are true.

What are the consequences for physical reality? We tend to think, based on our experience of the world, that large physical objects like books or astronauts can only be in one place at once and only a single fate can befall them. Quantum theory destroys this picture when we ask questions about the behaviour of sub-atomic particles, and complementarity appears to be an even more radical challenge to our intuition. It asks us to accept that there are two equally valid views of what happens to a big thing like an astronaut – you, for instance – freely falling towards a black hole. You are both spaghettified (inside) and vaporised (outside). This suggests that the relationship between the inside of a black hole and the outside is not the same as the relationship between 'here' and 'over there' according to our everyday experience. There is also a small but persistent fly in the ointment: Stephen Hawking's calculation, which says that the Hawking radiation contains no information. In the history of the subject, attacking this sharp, well-defined problem in Hawking's calculation turned out to be extremely fruitful in the quest to place complementarity on a rigorous footing, because the question is simple and easy to state: if the information is to come out, where did Stephen Hawking go wrong?

12

THE SOUND OF ONE
HAND CLAPPING

'Entanglement is iron to the classical world's bronze age.'
Michael Nielsen and Isaac Chuang[38]

Physicists love a paradox. Perhaps unusually, they spend their professional lives searching for situations that will cause their world view to collapse, because deeper understanding may grow from the rubble. Good scientists don't want their beliefs to be vindicated through research. They want research to generate new beliefs. The intellectual value of the black hole information paradox and the related affront to common sense delivered by black hole complementarity is that it forces physicists into a corner. For determinism to be preserved, there must be an error in Hawking's calculation. If there is no error, determinism must be sacrificed. Along either path, new insights await. Hawking's calculation, after all, rests on the very well understood foundations of quantum theory and general relativity.

What happens at the singularity of a black hole is beyond current understanding. According to general relativity, the singularity marks the end of time for anything unfortunate enough to meet it. It appears to be a place where matter ceases to exist. The inability to compute in the region of the singularity is a major unsolved problem in physics. Hawking radiation, on the other hand, is not a phenomenon that requires an understanding of physics close to the singularity. Hawking's calculation never strays beyond what physicists refer to as 'low-energy physics' – the (apparently) well understood domain of quantum physics and relativity in the near-horizon region. As we shall see, it turns out that our familiar low-energy laws do contain traces of the deeper theory of quantum gravity and these traces are made visible by studying Hawking radiation.

The first key insight came from understanding the unusual way Hawking radiation is produced and the constraints this places on the evaporation process – in particular, the fact that Hawking radiation is 'plucked' out of the quantum vacuum. Take another look at Figure 10.1, which shows the emergence of a pair of particles from the quantum vacuum. Because these two particles have their origin in the vacuum, they share certain properties with it. Most importantly, they are quantum entangled.

Of all the bizarre aspects of quantum physics, none is perhaps quite so bizarre as the notion of entanglement. According to Schrödinger, entanglement is the phenomenon that forces the quantum world's departure from classical thought, and it is the aspect of quantum physics that Albert Einstein referred to as 'spooky action at a distance'. Entanglement isn't considered so spooky today in the sense that it has become a part of our technology. It is the intellectual (and physical) resource that underpins the nascent quantum computers programmed and studied in many research laboratories. Entanglement is counter-intuitive, but it is also a tangible property of the world.

Entanglement has no counterpart in our everyday experience, which is why it appears counter-intuitive. Roughly speaking, it is a correlation between two or more things that is inexplicable using classical logic.* Entangled objects that are very far apart

---

* Correlations are commonplace in everyday life: the colour of your left sock is likely to be correlated with the colour of your right sock and living in Manchester is correlated with experiencing drizzle.

'feel' each other's influence instantaneously because they should really be viewed as a single connected system. This means that, underlying our everyday experience, there exists a more subtle, holistic world. There are electrons in your hand and electrons in the Andromeda Galaxy, separated by over 2 million light years, linked through quantum entanglement. This sounds like a near-mystical claim – spooky even. Crucially for the logical coherence of the world, however, these correlations cannot be exploited to send messages at faster-than-light speed, so don't get too excited. Nobody will be using quantum entanglement to build time machines. Having said that, these remarkable correlations are real.

Qubits

To explore entanglement, we'll introduce the notion of a quantum bit, or 'qubit'. An ordinary bit is like a switch; it can only have two values, which we might call 'on' and 'off' or 0 and 1. These familiar classical bits are the basis of all modern computing. Qubits are a far richer resource because they can be both 0 and 1 at the same time. Whenever we measure the value of a qubit it will return either a 0 or a 1, but beforehand it can be a mixture of both. In the jargon, we say that the qubit is in a linear superposition of 0 and 1. If you've heard of the famous Schrödinger's cat thought experiment, you'll be familiar with this idea. A cat is sealed in a box, and it has been arranged (using a convoluted experimental setup involving decaying atoms and a vial of poison) that the cat is both alive and dead if the box remains sealed. When the box is opened, the cat will be observed to be either alive or dead. This is treating the cat like a qubit – it can be both 0 and 1 until observed. We won't go into what constitutes an observation here, or why it may be appropriate to treat an object as large as a cat as a purely quantum system; for more detail you can read virtually any popular book on quantum mechanics, including our own, *The Quantum Universe*. All we need to know here is that qubits have a far richer structure than ordinary bits because they don't have to be either 0 or 1; they can be both 0 and 1 at the same time.

Paul Dirac introduced a powerful notation to represent qubits and quantum states in general. Let's consider a qubit which we'll label 'Q'. If it has a definite value of 1 then, in Dirac's notation, we write:

$$|Q\rangle = |1\rangle$$

If it has a definite value of 0 then we write:

$$|Q\rangle = |0\rangle$$

An example of a qubit with an equal chance of it returning 0 or 1 when read-out (observed) is:

$$|Q\rangle = \frac{1}{\sqrt{2}}\,|0\rangle + \frac{1}{\sqrt{2}}\,|1\rangle$$

This has no counterpart in classical computing logic. A qubit that will return 0 for 10 per cent of the time, and 1 for 90 per cent of the time, is:

$$|Q\rangle = \sqrt{\frac{1}{10}}\,|0\rangle + \sqrt{\frac{9}{10}}\,|1\rangle$$

This is how the quantum rules work. We are to square the numbers to get the probabilities. This state is 'mostly' 1 with a small mix of 0. It's worth emphasising that this qubit is not 'secretly' a 1 or a 0 and for some reason we don't know which. It really is both 0 and 1 at the same time. This is very counter-intuitive, but it's the way our Universe works.

Entanglement is a different but related idea. Imagine we have two qubits. If they are both 0, we could write their combined quantum state Q_2 as:

$$|Q_2\rangle = |0\rangle|0\rangle$$

where the first refers to the first qubit and the second refers to the second qubit. We could also imagine a state:

$$|Q\rangle = \frac{1}{\sqrt{2}}|0\rangle|1\rangle + \frac{1}{\sqrt{2}}|1\rangle|0\rangle$$

This is an entangled state.* There is a 50 per cent chance that the first qubit will have value 0 and the second qubit will have value 1, and a 50 per cent chance that the first qubit will have value 1 and the second qubit will have value 0. But notice that there is *no chance* that both qubits will be 0 or both qubits will be 1. A photon is an example of a physical system that behaves as a single qubit. It possesses a property known as spin, which can be either 0 or 1. The entangled 'Bell state' above can be realised as a system of two photons. States like this are routinely created in laboratories to study entanglement and for use in quantum cryptography and computing.

Let's put these qubits aside for the moment, and switch to a wonderful analogy developed by Paul Kwiat and Lucien Hardy known as the quantum kitchen.[39] Replace cakes for photons and change a few other words in what follows, and the story relates to real experiments that have been carried out in laboratories. The quantum kitchen is illustrated in Figure 12.1. The kitchen sits in the middle and two conveyor belts emerge from either side. Pairs

---

* This entangled state is an example of what is known as a 'Bell state', named after Northern Irish physicist, John Bell, who pioneered early studies into quantum entanglement.

of ovens travel along the conveyor belts and inside each oven is a cake that is baking as the oven moves. The cakes will be examined by Lucy (on the left) and Ricardo (on the right). The ovens can be opened halfway along, allowing the baking cakes to be observed. They will either have risen or not risen. At the end of the line, Ricardo and Lucy can make a different observation by eating the cakes. They will either taste good or bad. This is the experimental setup.

Lucy and Ricardo are going to encounter many pairs of cakes, and for each pair they will make a random choice as to whether to open the oven halfway along and look at the cake or to wait and taste the cake at the end. They are only allowed to make one observation each per pair of cakes; they can either open the oven halfway along and check whether their cake has risen, or they can taste their cake at the end, but not both. Their task is to record the results.

For the first pair of cakes, Lucy tastes her cake at the end of her conveyor and it tastes good. Ricardo does the same and finds that his cake tastes bad. For the second pair of cakes, Ricardo opens his oven halfway along and sees that the cake has risen. Lucy tastes her cake at the end and finds that it tastes good. And so on. After many cakes, they find:

1. Whenever Lucy's cake rose early, Ricardo's cake always tasted good.
2. Whenever Ricardo's cake rose early, Lucy's cake always tasted good.
3. In cases where both Lucy and Ricardo checked the ovens halfway along, 1/12 of the time both cakes had risen early.

If we use our common sense, honed by our experience of the world, these three observational facts lead us to infer that both cakes should taste good at least 1/12 of the time. We can infer this because:

1. when Ricardo and Lucy checked both ovens, they found that in 1/12 of the cases both cakes had risen, and
2. we know that when Ricardo's cake has risen, Lucy's tastes good and vice versa.

There is nothing surprising here, other than that they are rubbish bakers. However, here comes a shocking observation. In the quantum kitchen:

Both cakes never taste good.

How can that be? Using what seems like unimpeachable logic, we have concluded that in at least 1/12 of cases both cakes should taste good, and yet they never do.

It's fun to play around and try to work out what is wrong with this reasoning – what possible mechanism could be in play to produce these strange results? Physics students like to do this in the pub. Long ago, the authors spent an evening figuring out why attaching a rod to the Moon and tapping it in morse code could not violate the principles of relativity by sending signals faster than light.

There could be some mechanism whereby the only way to make a good-tasting cake is if the other person opens their oven at the midway point. Maybe the act of opening one oven midway causes a sound that shakes the other oven in just such a way as to make its cake taste good. Or perhaps a tiny, heat-resistant culinary gnome is sitting in each oven watching what is happening to the other oven who ensures that their cake tastes good if they see the

Figure 12.1. The quantum kitchen, from Paul Kwiat and Lucien Hardy's 'The Mystery of the Quantum Cakes'.

other oven door opened. Both these possibilities (which exploit a causal link) can be ruled out if we make the conveyor belts sufficiently long and fast-moving that no signal can travel between the ovens before the observations are made. The signal would need to depart after Lucy/Ricardo has decided to open their oven at the midway point and arrive at the other oven before Ricardo/Lucy tastes their cake. We can arrange things so that there isn't enough time for this to happen. Or perhaps the chef who makes the cake mixture somehow knows in advance the measurement choices that Ricardo and Lucy are going to make. The chef could then produce a bad mixture at just the right time. This possibility can be eliminated if Ricardo and Lucy make their random decisions after the ovens have left the kitchen. And so on. There is one logical possibility that could explain the results without resorting to quantum theory: every event in the Universe is predetermined, freewill does not exist and the results were baked in at the beginning of time. Putting that aside, we are left with quantum mechanics.

The results we've quoted above can be explained if the cakes are produced in an entangled quantum state. In this case, the quantum state of the cakes is such that both cakes can never taste good. This is like the situation we encountered earlier with our entangled system of qubits; there was no chance that both qubits could be 0 or 1.

Here is a quantum state for the cakes that reproduces the results obtained by Lucy and Ricardo:

$$|Q\rangle = \frac{1}{\sqrt{3}} \left(|B_L\rangle|B_R\rangle - |B_L\rangle|G_R\rangle - |G_L\rangle|B_R\rangle \right)$$

where B and G mean 'tastes bad' and 'tastes good' respectively and the subscripts refer to Lucy and Ricardo's cakes. This is a more complicated state than we've seen before, but you can see that, because there is no $|G_L\rangle|G_R\rangle$ term, both cakes never taste good. You can also see that both cakes taste bad one third of the time. To explain the numerical results associated with opening the ovens and observing whether the cakes have risen, we need a bit more quantum theory that won't be necessary for what follows, but it is interesting, so we've moved the discussion to Box 12.1.

BOX 12.1. More from the quantum kitchen

You might be wondering where the 'risen' or 'not risen' measurements fit into the quantum state of the cakes. To reproduce the results obtained by Lucy and Ricardo, the 'tastes bad' and 'tastes good' states of the cakes are given by:

$$|B\rangle = \frac{1}{\sqrt{2}}(|N\rangle + |R\rangle)$$

$$|G\rangle = \frac{1}{\sqrt{2}}(|R\rangle + |N\rangle)$$

where R and N mean 'risen' and 'not risen'. How should we interpret these states? Let's take $|B\rangle$ as an example. If a cake is observed to taste bad, then this means it must be in state $|B\rangle$. If we now make a subsequent observation and ask whether it has risen or not, there is a 50 per cent chance it will not have risen, because $1/\sqrt{2}$ squares to ½. If you fancy a little bit of mathematics, you can substitute these expressions for $|B\rangle$ and $|G\rangle$ into the state $|Q\rangle$ to ascertain the coefficient of the $|R_L\rangle|R_R\rangle$ term. You should find it's $-1/\sqrt{12}$ which gives a probability of $1/12$ that both cakes will have risen. With less effort, you can also see that there is no $|R_L\rangle|B_R\rangle$ and $|B_L\rangle|R_R\rangle$ piece in $|Q\rangle$, which explains facts 1 and 2. Our quantum chef is responsible for preparing the cakes in these very specific states.

For our purposes, the most important feature of the quantum kitchen is that, because the cakes are in an entangled state, they do not possess the qualities of 'tastes good', 'tastes bad', 'risen' or 'nor risen' independently. Rather the whole two-cake system is produced by the quantum kitchen in a state that mixes all these possible measurement outcomes with correlations that give the results we quoted above. And yet the quantum state is also such that in every case, each individual cake has the potential to taste good or bad, or to rise or not rise, when it leaves the kitchen. To reiterate a very important point, the probabilities observed by Lucy and Ricardo are not the result of a lack of knowledge about the state of the quantum cake system. They can know the state and yet before a measurement is made they cannot know whether an individual cake will taste good or bad, or is risen or not because each cake is all of these, until it's observed.*

Where does the information about the correlations in the entangled state reside? It is not stored locally by each individual cake. Returning to our simple two-qubit system:

$$|Q_2\rangle = \frac{1}{\sqrt{2}}\,|0\rangle|1\rangle + \frac{1}{\sqrt{2}}\,|1\rangle|0\rangle$$

a measurement of one of the qubits will reveal 0 or 1 with equal probability. That's just like tossing a coin. Half the time it will come up heads and half the time it will come up tails. There is no information in that; it's completely random. And yet, if the coins were quantum coins entangled in this way, if one of the coins

---

* It is an important feature of quantum mechanics that this 'link' between the cakes cannot be used to transmit information faster than light. We could ask for the probability that Lucy finds a good-tasting cake – something that Lucy can measure independent of Ricardo. We will discover that the odds Lucy observes do not depend upon Ricardo's measurements, which means that Ricardo cannot use his choice of measurement to transmit a message to Lucy. Even though their observations are correlated, it is not a correlation that can be used to transmit information.

came up heads, we would know for certain that the other came up tails. This would be true even if the coins were on opposite sides of the Universe. There is information stored in the state that prevents both the coins from ever coming up heads or tails at the same time, but it's not stored in a way that is familiar to us. In a book, information is stored locally on each page and as we read more pages, the story gradually unfolds. In a quantum book each individual page would be gibberish and the story would reside in the correlations between the pages. We'd therefore have to get through a good portion of the book before we gained any insight into the story at all. No correlations, and therefore no information, would be discernible from a single page. The lack of information stored in a small part of a large, entangled quantum system is a very important property, which is central to the black hole information paradox.

Entanglement and evaporating black holes

The quantum vacuum is anything but empty. It is also heavily entangled. The extent to which the vacuum is entangled is captured by the remarkable Reeh–Schlieder theorem, which states that it is possible to operate on some small region of the vacuum in such a way that anything can be created anywhere in the Universe. This phenomenal conjuring trick is theoretically possible because the vacuum is inexorably entangled. The outlandish nature of the theorem is diminished ever-so-slightly by the fact that the local operation required is not something that we could ever perform, which is a shame. Nevertheless, the point remains that the vacuum has this encoding within it. Importantly for us, Hawking radiation is a product of this vacuum entanglement.

Figure 12.2 illustrates the process of black hole evaporation by the emission of Hawking radiation. The dotted lines indicate pairs of Hawking particles that are entangled because of their origin in the quantum vacuum. Since the Hawking pairs necessarily strad-

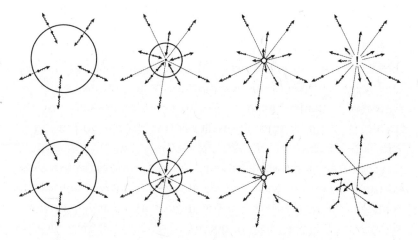

Figure 12.2. Entanglement between Hawking radiation and a black
hole. At the top, how things proceed in the original Hawking
calculation, where information is lost. Entanglement steadily increases
between the radiation and the hole which causes a puzzle when the hole
finally disappears (top right). At the bottom, how things will go if
information does not get lost. The entanglement slowly transfers
from being between the hole and the radiation to being entirely
within the radiation.

dle the event horizon, the entanglement can be pictured as being
between the Hawking radiation exterior to the black hole and the
black hole itself. As time passes, more and more Hawking radiation emerges, and more and more radiation particles become
entangled with the black hole. But the black hole is shrinking.
Eventually it disappears, and we have a problem, because the
thing the Hawking radiation was entangled with has disappeared.
The Hawking radiation left behind is like the sound of one hand
clapping.

What are the consequences of the disappearance of entanglement? As we've seen, an entangled system has a rich structure that
encodes information in the correlations across the system. That
information must necessarily be lost if the entanglement is broken,
which violates the rules of quantum mechanics and the basic prin-

ciple of determinism.* This is the essence of the black hole information paradox.

In order to dodge this unpalatable scenario, we might appeal to the fact that we do not understand what happens during the very final stages of black hole evaporation. When the hole is large, Hawking's calculations are expected to be reliable because the spacetime in the vicinity of the horizon is not curved too much. This isn't the case when the hole gets very small, just before it disappears. In these final moments, the spacetime curvature at the horizon becomes so great that we should not expect to be able to apply general relativity and quantum theory to make predictions. We are entering the realm of quantum gravity and, given that we don't have such a theory, we might reasonably claim that all bets are off. Perhaps therefore, it would be wise to take a more cautious view and say that the information paradox may be resolved by some as-yet-undiscovered physics.

In 1993, North American physicist Don Page established[40] that this line of reasoning is at fault and that appealing to unknown physics late in the black hole's life doesn't solve the problem because a paradox arises much earlier in the evaporation process, when the hole is middle-aged. Think of the black hole and the Hawking radiation together as a single entangled system that we are dividing into two; a more complicated version of the quantum kitchen. The black hole is one cake and the radiation is the other, and they are entangled. As more Hawking radiation is emitted, the hole shrinks with the result that more and more (radiation) is becoming entangled with less and less (the shrinking hole). There comes a point when the shrunken black hole no longer has the

---

* In quantum mechanics, we use the word determinism to refer to the fact that we can predict the future state of a system if we know its prior state. However, since quantum mechanics is inherently random, knowing the state does not also mean we know the results of experiments. In this regard, quantum mechanics is not deterministic. The more precise terminology is to say that quantum states undergo 'unitary' evolution.

capacity to support the entanglement with the emitted radiation and, as Don Page realised, that will happen when the black hole is middle-aged.

We can illustrate Page's reasoning with an analogy. Imagine a jigsaw puzzle made up of square pieces and imagine that the completed puzzle is set out on a table. The completed puzzle contains a large amount of information – the picture on the jigsaw. Imagine also a second empty table, onto which we will transfer pieces randomly selected from the jigsaw. The completed puzzle on the first table is like a black hole before it emits any Hawking radiation, and the empty table is the exterior of the black hole.

We now take a piece out of the completed puzzle at random and move it to the empty table, followed by another piece and then another. The pieces on the second table are like the Hawking particles emitted by the black hole. There are now three pieces on the previously empty table. These pieces are unlikely to reveal the jigsaw picture to us. If we focus only on these three pieces, we have no inkling that they are a part of a larger, more correlated and information-rich system.

The entropy of the pieces on the second table counts the number of ways we can arrange the pieces on the table if we pay no attention to whether the pieces fit together or not. This is like computing the Boltzmann entropy of a gas by counting arrangements of atoms (we are ignorant of the precise details). We will refer to this as the thermal entropy.*

At first, when only a few pieces have been transferred, the entropy of table two increases with each piece we transfer because there are more pieces and we can put them anywhere we want.

---

* For example, if the jigsaw is a 3 x 3 puzzle with square pieces, where the pieces occupy one of 9 possible positions on a grid, the number of possible arrangements is $9 \times 8 \times 7 \times 6 \times 5 \times 4 \times 3 \times 2 \times 1 \times 4^9 = 95{,}126{,}814{,}720$. The thermal entropy of the jigsaw is the logarithm of this number. The factor of 4^9 is because each piece can be placed in one of four different orientations.

However, when a large enough number of pieces have been transferred to the second table, the pieces start to fit together. At some
point, therefore, adding more pieces does not increase the number
of ways we can arrange the pieces on the second table. Rather,
adding more pieces leads to fewer possible arrangements as we
start to see the bigger picture.

To make this more quantitative, we can introduce a new kind
of entropy called the entanglement entropy. This entropy counts
the number of possible arrangements of the pieces accounting for
the fact that some pieces fit together. When only a few pieces have
been transferred, the entanglement entropy is equal to the thermal
entropy because none are likely to fit together. As more pieces are
transferred, however, the entanglement entropy will eventually
start to reduce because more and more pieces will fit together,
restricting the number of possible arrangements. The thermal
entropy, on the other hand, will continue to rise because it is
concerned only with the number of pieces on the table.

The reason why this new quantity is called the 'entanglement
entropy' is that it is a measure of how entangled the two jigsaws
are. It is zero when no pieces have been transferred and it starts to

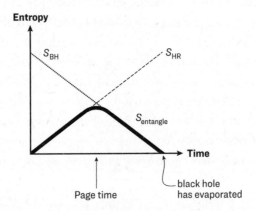

Figure 12.3. The Page curve (black curve). Also shown is the
Bekenstein–Hawking result for the entropy of a black hole (dotted line)
and Hawking's result for the entropy of the radiation (dashed line).

increase as pieces get transferred. All the information contained in the jigsaw is still there, but it is now starting to become shared between the two tables. At some point, the entanglement entropy starts to fall again as the completed jigsaw begins to emerge on the second table. Information is now starting to appear on the second table and the amount of shared (entangled) information is falling. The two parts of the jigsaw are most entangled with each other when about half the pieces have been transferred, which is when the entanglement entropy is at its maximum value. For the jigsaw, therefore, the entanglement entropy starts from zero, rises to a maximum when both tables contain roughly the same number of pieces, and then falls back to zero again. We have sketched this in Figure 12.3.

This simple jigsaw analogy provides a way to understand Page's reasoning. The first table is analogous to the black hole and the second table is analogous to the emitted Hawking radiation. If information is conserved in black hole evaporation, we have just learnt that the entanglement entropy between the black hole and the radiation should first rise and then fall, as illustrated in the Page curve of Figure 12.3. The Page time is the time when the entanglement entropy stops rising and begins to fall. It marks the time when the correlations between the Hawking particles start to carry a significant part of the total information content of the original system.

We have also drawn the thermal entropy of the black hole, which gradually falls to zero as the black hole evaporates away, and an ever-rising thermal entropy for the Hawking radiation. This ever-rising curve is the result of Hawking's original calculation, which appears to show that there are never any quantum correlations in the radiation. Page's powerful point is that an information-conserving evaporation process must follow the Page curve and *not* Hawking's ever-rising curve. And, crucially, the difference between the two curves manifests itself at the Page time, which is when the black hole is not too old and therefore quantum theory and general relativity should both be valid. Viewed this way, solving the information paradox is tantamount to understanding which curve is correct: the Page curve (information comes out) or Hawking's original curve (information does not come out).

The Page time can also be thought of as the time at which we could begin to decode information contained in the Hawking radiation (if the information comes out). If the radiation is thermal, as it is to a good approximation before the Page time, then its entanglement entropy is equal to its thermal entropy. In this case, no correlations are visible and no information is contained in the radiation. After the Page time, correlations appear and the radiation becomes increasingly information rich. This is the situation we previously described for a quantum book. The initial pages are complete gibberish because the story is encoded in the correlations between the pages. It's only when we get more than halfway through the book that we can begin to identify the correlations and decipher the meaning.

The fact that we have to wait until the Page time for any correlations to appear leads to another quite bamboozling idea. Because the Page time is roughly halfway through the lifetime of a black hole, which may be in excess of 10^{100} years, we are claiming that correlations appear between particles whose emission is separated by 10^{100} years. This illustrates the strangeness of quantum entanglement and perhaps hints that space and time may not be what they seem.

The conclusion we are forced into is that if information is to be conserved, some correction to Hawking's original calculation must be present not much later than the Page time, which is approximately halfway through the black hole's lifetime. But to reiterate a crucial point: at the Page time we expect both general relativity and quantum theory to be perfectly adequate in the near-horizon region; we would not expect to need presently unknown physics. And yet Hawking's calculation, based on quantum theory and general relativity, diverges from expectations if we believe that information should be conserved in black hole evaporation. The challenge is now clear. If we are to show that information is not destroyed by black holes, we must calculate the Page curve.

We now find ourselves in the position of the theoretical physics community around the turn of the millennium. Don Page had laid down the gauntlet because the Page curve should be calculable using the known laws of physics. But the state-of-the-art calculation at the time was Hawking's, which did not follow the Page curve. There was also complementarity, the crazy idea we met in the previous chapter, which lacked a convincing proof. Complementarity offers a solution to the information paradox in the sense that no information ever actually falls across the horizon and into the black hole as viewed from the outside – but we are also asked to believe that there is another point of view in which information does fall in, and that both points of view are equally valid. Around this time, a bold new idea surfaced that was crucial in convincing many that Hawking must be wrong, and that complementarity has substance. The key idea? The world is a hologram.

13

THE WORLD AS A HOLOGRAM

'Nobody has the slightest idea what is going on.'

Joseph Polchinski

The entropy of a black hole is proportional to its area, which suggests that all the information concerning the stuff that fell into the hole is encoded in tiny bits spread over the surface of the horizon. In time, those bits break free and end up as correlated Hawking particles. These correlations – quantum entanglement in the radiation – encode the information about the stuff that fell in.* From the perspective of someone freely falling through the horizon, they feel nothing and are oblivious to this magical encoding. Moreover, their fate is to be both spaghettified in the

---

* This is decidedly not what happens in Hawking's calculation, which suggests that the Hawking particles are uncorrelated and therefore carry no information, violating a fundamental principle of quantum mechanics. In Hawking's calculation, the particles come out of a data-less vacuum (in the words of theoretical physicist Samir Mathur). As a result, the entanglement entropy of the Hawking radiation increases indefinitely and never turns around (as it must to accord with Don Page's curve) because empty space is effectively providing an infinite reservoir of entanglement. It is radical to say that the information encoded on the horizon gets transferred to the outgoing Hawking radiation and incumbent on us to find the theory that explains how that comes about. Providing this explanation is what will constitute a full resolution to the information paradox. Complementarity does not answer the question directly – it supposes that some as-yet-unknown dynamics of the hot region near the horizon leads to large corrections to Hawking's calculation.

singularity (from their own perspective) and burnt up on the horizon (from an outsider's perspective). But that is no problem for the laws of Nature because no observer can be present at both events. This is the essence of the black hole complementarity resolution to the black hole information paradox. Is it nonsense?

Today, the evidence is strongly in support of the conclusion that complementarity is not nonsense, but the implication is even more shocking. Complementarity is telling us that what happens inside the horizon is as valid a picture of physical reality as what happens outside. The two pictures are equivalent descriptions of the same physics. The inside of a black hole, in other words, is somehow 'the same' as the outside. This idea has become known as the holographic principle.

At first sight there might seem to be no need to be so radical as to invoke the holographic principle because we could imagine that, as something falls through the horizon, it is secretly copied. One copy continues to fall into the singularity to be spaghettified and the other copy gets burnt up on the horizon and encoded in the Hawking radiation. Radical as it may be, copying on the horizon does seem less radical than invoking the idea that the interior is a hologram. There is a serious flaw in this logic though – the laws of quantum physics forbid it. The 'no cloning theorem' says that it is not possible to make an identical copy of some unknown quantum state, and in Box 13.1 we sketch a proof.

BOX 13.1. No cloning

Suppose we have a cloning machine that can duplicate an unknown qubit. Specifically, this machine takes a $|0\rangle$ and turns it into $|0\rangle |0\rangle$ and likewise it turns a $|1\rangle$ into a $|1\rangle |1\rangle$. What would it do to the following qubit: $|Q\rangle = 1/\sqrt{2}(|0\rangle + |1\rangle)$? This qubit is a 50–50 mix of $|0\rangle$ and $|1\rangle$ and our cloning machine would turn it into $1/\sqrt{2}(|0\rangle|0\rangle + |1\rangle|1\rangle)$. But this two-qubit state is not $|Q\rangle |Q\rangle$. In other words, our machine is not a cloning machine after all.

With cloning ruled out, it appears that we are left with holography if we want to respect the foundations of both quantum
theory and general relativity. There is, however, another possibility: the black hole has no interior. This radical solution would
mean general relativity is wrong because nothing can fall into a
black hole, which is a gross violation of the Equivalence Principle.
This outrage to Einstein was taken very seriously after a 2013
paper by Ahmed Almheiri, Donald Marolf, Joseph Polchinski and
James Sully, provocatively titled 'Black Holes: Complementarity
or Firewalls?'[41] The AMPS paper, as it has since become known,
found what appeared to be a fatal flaw in the complementarity
idea. This led to the proposal that a black hole has no inside, and
that anyone unfortunate enough to reach the horizon of a black
hole would be burnt up in a wall of fire, even from their own
point of view.

Firewalls

In Chapter 12, we imagined loitering outside a black hole and
collecting the evidence of an astronaut's fate as they get burnt up
on the horizon. We would then jump into the hole to confront
the same astronaut with their own ashes. This would generate a
contradiction, because the astronaut would have both burnt up
and not burnt up *from their own perspective*. We explained that
this contradiction is avoided because the astronaut will have
reached the singularity before we are able to catch them. The more
precise version of this scenario involves thinking of qubits and
cloning. We can imagine throwing a bunch of qubits into the hole
and then trying to determine those bits by collecting the Hawking
radiation and processing it so that we have effectively obtained a
copy of the original qubits we threw in. To be consistent with the
no cloning theorem, it should not be possible to do that and then
jump into the hole and meet up with the original qubits. From
our understanding of the Page curve, we know that if the black

hole is younger than the Page time, not much information will have emerged and we would need to wait a long time (a silly understatement for young, solar mass black holes) before we could obtain a copy of the original qubits. There should be no contradiction, therefore, for a young black hole.

The situation after the Page time, when the black hole is middle-aged or older, is rather more subtle. In that case, the black hole acts more like a mirror and spits the bits back out again almost immediately. That discovery was made in 2007 by Patrick Hayden and John Preskill.[42] Surprisingly, however, it turns out that the time delay is still (just) sufficient to prevent a violation of the no cloning theorem. All appears well in the complementarity camp, but AMPS seemed to throw a spanner into the works by coming up with a similar thought experiment. Their scenario, however, cannot be so easily reconciled with complementarity.

In the last chapter, we saw that if information is to be transferred into the Hawking radiation after the Page time, the Hawking particles must gradually become more and more entangled with each other. This is illustrated in the lower half of Figure 12.2. However, the Hawking radiation is produced as entangled

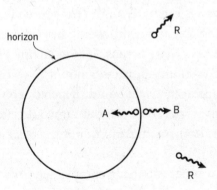

Figure 13.1. Illustrating the firewall. For an old black hole, the Hawking particles emitted when the hole was young, R, are entangled with the recently emitted Hawking particle B, which is also entangled with the interior particle A.

pairs, as illustrated in the upper half of Figure 12.2. And here is the problem: because the Hawking pairs are entangled with each other they cannot be entangled with anything else. This is known as the 'monogamy of entanglement', and it is another fundamental property of quantum mechanics.

Figure 13.1 illustrates the problem this causes for a black hole older than the Page time. Imagine two observers, Alice and Bob. Bob, sitting outside of the black hole, collects the Hawking radiation. He processes R, the radiation emitted early in the life of the hole (before the Page time), and distils it into a single qubit.* Now, if information is to be conserved, this qubit is highly likely to be entangled with a Hawking particle B that is emitted late on in the life of the hole – this is why the Page curve goes to zero when the hole finally disappears. Bob therefore concludes that B and R are entangled.

Alice is a freely falling observer who crosses the horizon after the Page time. She will confirm that particle B is entangled with the other half of its Hawking pair, labelled A. To avoid violating the monogamy of entanglement, while still permitting the information to come out in the Hawking radiation,† we could suppose that Alice does not confirm A and B to be entangled. This might sound innocuous, but it is not. The consequences of simply removing entanglement like this would be very dramatic: it would create a wall of fire. That's because it costs energy to destroy entanglement in the vacuum – it is tantamount to tearing open empty space. The resulting firewall wouldn't merely prevent Alice from entering the interior of the black hole, it would effectively destroy space inside the horizon. The interior of the black hole would not exist.

One might wonder whether a complementarity-style argument might still save the day. Maybe Alice could observe that A is

---

* By some ingenious and complex process that we do not need to know about here.
† By which we mean Bob still detects the entanglement between B and R.

entangled with B and Bob could observe that B is entangled with R with no contradiction because they can never meet to confirm their observations. This is not the case though because there is plenty of time for Bob to confirm that he sees entanglement, and then to dive across the horizon to compare notes with Alice, who would have confirmed that she too sees entanglement across the horizon by the very act of crossing it.

With this chain of reasoning, AMPS appeared to have discovered a genuine contradiction which calls into question the existence of the interior of a black hole. This is possibly what provoked Joseph Polchinski to utter the words that open this chapter. The basic problem can be traced to the fact that complementarity appears to be asking for too much entanglement in order to conserve information as the black hole evaporates and to simultaneously preserve the integrity of the quantum vacuum across the horizon. Complementarity requires the black hole and the Hawking radiation to be in an impossible quantum state after the Page time, by demanding they encode more information than the system can physically support.

In the conclusions to their paper, AMPS mention another way of seeing this information storage limit in action. Why don't similar arguments imply that firewalls should appear at the Rindler horizons we met in Chapter 3 for accelerated observers? The answer is that Rindler horizons have infinite area and entropy, unlike a black hole horizon, and therefore 'their quantum memory never fills'. Rindler horizons, in other words, never evolve to become old and can always support any information demands made of them to preserve the integrity of the vacuum.

Holography offers a means to save the Equivalence Principle and render the horizon safe, while also saving quantum mechanics and preserving information. The basic idea is that since the interior of the black hole is dual to the exterior; the early Hawking radiation, R, and the interior particles, A, are really the same thing. Crazy as it sounds, this is the way the firewall problem is avoided in holography. In processing the early Hawking radiation, R, to check its entanglement with B, Bob inadvertently destroys the entanglement with A. This creates a kind of mini-firewall that is just violent enough to prevent Alice from measuring entanglement between A and B, but not so violent that it destroys the interior of the black hole.

The world as a hologram

Spacetime holography was first presented by Gerard 't Hooft in 1993, and further developed a year later by Leonard Susskind. They presented it as an integral part of their black hole complementarity idea but stressed that it probably ought to be of more general applicability. That's to say, holography should be a universal feature of Nature, regardless of the black hole issues that originally motivated it. The holographic principle as currently understood even goes so far as to suppose that the entire world as we perceive it is a hologram.[43]

A hologram as conventionally understood is a representation of a three-dimensional object constructed from information stored on a two-dimensional screen. If you've ever seen a hologram, you'll know that they can look remarkably real. You can walk around them and view them from all angles as if they were the real three-dimensional objects themselves. Now imagine a perfect hologram. What sort of a thing would that be? A hologram would be perfect if every bit of information necessary to reconstruct the three-dimensional object was also encoded on the two-dimensional screen that carries the holographic data. This is reminiscent of the Bekenstein entropy of a black hole, which says that the information content of the black hole can be computed by considering only the two-dimensional surface of the event horizon.

Now, as we discussed in Chapter 9, a black hole has the largest possible information density of any object and, since the information it stores is given by the surface area of the horizon, it follows that there can be no more information inside *any* region of space than can be encoded on the boundary of that region. This realisation led 't Hooft and Susskind to argue that the information content of any region of space is encoded on the boundary to that region. The reason we discovered this first by thinking about black holes is that the boundary is exposed for anyone hovering outside the black hole to explore, in the form of the hot membrane close to the event horizon. In everyday life, well away from any black holes, it is rather less obvious how we could gain access to this holographically encoded information since we cannot 'cut out a piece of empty space' to reveal that the information in the interior is encoded on its surface.

Holography, then, is a perfect example of complementarity in action. There are two entirely equivalent descriptions of anything and everything, and this is an essential feature of all of Nature and not just black holes. Black holes are the Rosetta Stone, which has introduced us to a new language; an entirely different yet perfectly equivalent description of physical reality. One description resides on the boundary of any given region of space, and the other resides more conventionally in the space internal to the boundary. The implication is that our experience and existence can be described with absolute fidelity in terms of information stored on a distant boundary, the nature of which we do not yet understand. This sounds utterly bonkers, but clinching evidence supporting the idea comes from the most highly cited high-energy physics paper of all time.

Maldacena's world

Scientific citations count how many times a research paper has been referred to in the literature. Naturally enough, the most important papers tend to get the most citations. Ranked number

13 of all time* is Stephen Hawking's 1975 paper, 'Particle Creation by Black Holes'. The discovery of dark energy is up there, with the two key papers reporting the evidence ranked 3 and 4, and the papers announcing the discovery of the Higgs boson at the Large Hadron Collider are ranked 6 and 7. Top-ranked of all time is a paper written in 1997 by Argentinian physicist, Juan Maldacena titled, 'The Large N Limit of Superconformal Field Theories and Supergravity'.[44] With almost 18,000 citations to date, it is the paper that has, more than any other, changed the face of theoretical physics over the past 25 years. It is also the paper that provides the strongest evidence supporting the idea that the holographic principle is true.

The universe Maldacena considered is not the one we live in, but that's fine. It is common for physicists to build models of the world with some simplifying features. The real world is complicated, and it's often useful to do calculations in a pretend world in which things are simpler. The skill is to pick a simple world which delivers enhanced understanding while not being too unrealistic. Engineers make simplifying assumptions when designing things like aircraft and bridges, notwithstanding the fact that the stakes are rather higher. Importantly, Maldacena's world was not specifically chosen because it supports holography. Rather holography was a feature that popped out of the mathematics.

We can capture the essence of Maldacena's work by considering a two-dimensional toy universe.† The space of the toy universe does not have the geometry of ordinary flat space; rather its geometry is hyperbolic. Figure 13.2 shows a beautiful

---

* These citation statistics come from the iNSPIRE database (inspirehep.net) which is run by a collaboration of the world's leading research laboratories and measures citations in the field of 'high-energy physics'.

† Maldacena's original calculation was in string theory and involved a ten-dimensional spacetime with five curled-up space dimensions, leaving a five-dimensional hyperbolic space with a four-dimensional boundary. Since 1997, there have been many other examples of the holographic principle involving fewer spacetime dimensions.

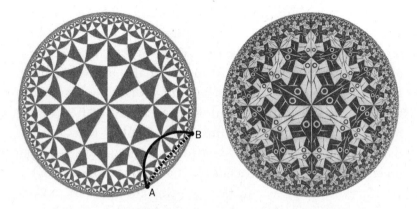

Figure 13.2. The Poincaré disk projection of a two-dimensional hyperbolic space. Despite appearances, the solid line from A to B on the left figure is shorter than the dashed one. You can see this by counting triangles. On the right is M. C. Escher's *Circle Limit I*. All the fish are of the same size and shape, and the lines are shortest lines. The patterns provide us with a visual representation of the metric of the space (we can ascertain distances by counting fish, or triangles) just like the squares on graph paper illustrate the metric of Euclidean space.

representation of two-dimensional hyperbolic space, known as the Poincaré disk. This projection was widely employed by the Dutch artist M. C. Escher, and we also include his well-known *Circle Limit I*. Rather like a Penrose diagram, the space represented in these projections is infinite and there is a great deal of distortion to bring infinity to a finite place on the page. Escher's fish, for example, are all the same size as they tile infinite hyperbolic space. They appear smaller towards the edge of the disk, which represents infinity, because we are shrinking down space as we head outwards from the centre. The Poincaré disk projection is also a conformal projection, which means that the shapes of small things are faithfully reproduced (for example, the fish-eyes are always circular).

Now let's add time into the mix. Figure 13.3 is a stack of Poincaré disks, one for each time slice (though we have drawn only two). Time runs upwards from the bottom of the cylinder. This spacetime is known as 'Anti-de Sitter spacetime', or AdS for short. For what follows, it helps to think of the AdS cylinder as being akin to a tin-can, with a boundary and an interior. Maldacena triggered an avalanche of understanding by showing that a particular theory with no gravity, defined entirely on the boundary of the cylinder, is precisely equivalent to an entirely different theory, with gravity, defined in the interior spacetime. In other words, the interior is a holographic projection of the boundary. By writing down equations proving the exact one-to-one correspondence for this model universe, Maldacena provided the first concrete realisation of the holographic principle.

To appreciate the key ideas, we don't need to know the details of what has become known as the AdS/CFT correspondence. The CFT acronym means 'conformal field theory', which refers to a class of quantum field theories that are similar to the ones used to build models of particle physics.* It refers to a quantum theory, complete with particles and entanglement and a vacuum state. The quantum theory describes a physical system located entirely on the boundary of the cylinder. If you'd like a picture, think of a gas of particles moving around.

When the quantum system on the boundary is in a pure vacuum state, which means there are no particles, the interior spacetime is just AdS. Now imagine creating particles on the boundary to make a gas. Astonishingly, a black hole appears in the interior spacetime. This is illustrated in Figure 13.3, where the formation and evaporation of the black hole in the interior has a dual description in terms of the gravity-free theory existing on the boundary of the cylinder. Gravity thus emerges as a result of the quantum mechanics of a system on the boundary.

---

* The particular CFT that Maldacena originally considered is similar to QCD, the theory describing the strong interactions between quarks and gluons, and that similarity has been exploited with some success to make predictions in QCD using the dual gravity theory.

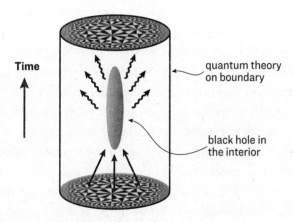

Figure 13.3. The Penrose diagram of Anti-de Sitter spacetime in the case of two space dimensions. The cylinder is infinitely long and the boundary is timelike. Some collapsing matter (at the bottom) makes a black hole that subsequently evaporates by emitting Hawking particles (at the top). The holography idea is that the formation and evaporation of the hole can be described using a gravity-free quantum theory defined on the boundary.

We might ask which of these two descriptions is the real one. Is there really a black hole or is it just a hologram of the boundary physics? Or maybe the opposite is true, and the boundary physics is not real and is just a clever way to describe the black hole. Maybe trying to figure out what is 'really' true is to fall into a trap that has long plagued physicists because it leads to navel gazing without revealing deeper insight. There are plenty of people in the world who can perform that function, and too few physicists, so perhaps we should restrict ourselves to explaining natural phenomena and leave questions of ultimate truth to others. Rather, the holographic principle can be viewed as a realisation of complementarity. There are two equivalent descriptions of the world, and because they are equivalent there will be no contradictions: What is true in one will be true in the other. This is the power of the holographic principle, and Maldacena discovered a precise, mathematical realisation of it.

This technique of mapping a problem in quantum physics to an equivalent problem in gravity has proved to be very successful over the past 25 years. Many cases have been found where complicated problems on one side of the correspondence have been answered using methods from the other side. Viewed this way, Maldacena discovered a practical tool that we are learning to use to solve interesting problems in one area of physics using techniques from what superficially looks to be a totally different area. This is one reason why Maldacena's paper has so many citations; it is very useful. It is also profound, and it answers the question of whether or not information is lost in black hole evaporation.

Figure 13.3 illustrates how the AdS/CFT correspondence demonstrates that information must come out of a black hole. Initially there was no black hole (the bottom part of the cylinder) – just a bunch of stuff collapsing under gravity. The black hole forms and then evaporates away leaving (at the top of the cylinder) a bunch of Hawking particles. Now focus on the dual description. This side of the correspondence says that this whole process can be described by a gas of particles evolving according to the ordinary rules of quantum mechanics on the boundary with no gravity. Because there is a precise one to one correspondence between the boundary theory and the interior theory, if information is conserved in one it must be conserved in the other. Crucially, the boundary theory is a pure quantum theory, which means that information is necessarily conserved. It must therefore be conserved by gravitational processes in the interior – in this case during the formation and evaporation of a black hole. This is what convinced Stephen Hawking to concede his bet with Kip Thorne and John Preskill and accept that information really does emerge from black holes in our Universe. He was persuaded by Maldacena's AdS/CFT paper.

14

ISLANDS IN THE STREAM

'By discovering the AdS/CFT correspondence, Maldacena definitively answered the question of whether information can escape from a black hole. It can. However ... we also need to understand what is wrong with the Hawking calculation.'

Geoffrey Penington[45]

What is it that makes a quantum theory on the boundary spacetime able to encode phenomena in the interior? How does holography work? Remarkably, and as we shall see in this chapter, it is as if the interior space is fabricated by quantum entanglement on the boundary. In other words, current research appears to be stumbling across the idea that space is not fundamental but rather something that emerges out of quantum theory: the quantum gravity puzzle may end up being resolved in favour of quantum mechanics with gravity emerging out of that.

In Chapter 6 we met the maximally extended Schwarzschild spacetime, which can be interpreted as representing two universes connected by a wormhole. We noted that, sadly, large traversable wormholes of the sort beloved by science fiction writers do not reside inside real black holes because the interior contains matter from the collapsing star. We did, however, say that 'microscopic wormholes could be part of the structure of spacetime'. It is now time for us to follow that thread.

Figure 14.1 shows the Penrose diagram of an eternal black hole that is very similar to the maximally extended Schwarzschild black hole we explored in Chapter 6 (see Figure 6.2). The difference is that this black hole is sitting in AdS spacetime. One might ask why we don't focus on a universe more like ours rather than on an AdS universe. The answer is that we would if we could, but we don't know how to yet, and Maldacena's AdS/CFT correspondence is the most well understood model we have to hand. The majority of experts in the field at the time of writing believe that the underlying ideas should also be valid in our Universe.

The upper and lower triangles represent the interior of the black hole,* bounded by the event horizon and the singularities in the future and the past. The edges of the diagram, labelled L and R, are the boundaries of the AdS spacetime. Just as for the Schwarzschild case, the left and right triangular regions are the entirety of the spacetime outside the black hole, and they are linked by a wormhole. The AdS/CFT correspondence tells us that we can describe the interior of this Penrose diagram by two quantum field theories (CFTs†) located on the left and right boundaries. In the jargon, we say that the interior spacetime is the holographic dual of these two quantum theories.

Now here is the big deal: the two CFTs must be maximally entangled with each other to describe this spacetime. If the two CFTs were not entangled, there would be no wormhole. Instead, there would be two disconnected black holes in two separate universes. The wormhole appears when we allow the two quantum theories to become entangled. In other words, entanglement builds the wormhole connecting the two universes together: quantum entanglement and wormholes go hand in hand. This is

---

* Actually, the bottom triangle is the interior of a white hole, as in Chapter 6.

† We are going to use the acronym CFT quite a bit in what follows. You can think of this simply as a gas of particles with no gravity acting.

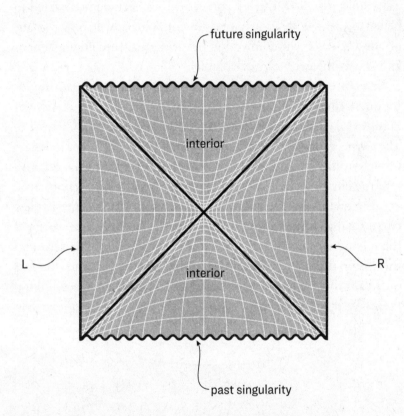

Figure 14.1. A two-sided black hole.

a profoundly important connection between quantum theory and gravity.

To explore this connection further, we'll introduce one of the central ideas to have emerged from holography: the Ryu–Takayanagi conjecture.[46] Discovered by Shinsei Ryu and Tadashi Takayanagi in 2006, the RT conjecture has been demonstrated to be correct in a wide variety of different scenarios. It is important because it makes the connection between quantum entanglement and spacetime geometry calculable.

In Figure 14.2 we have drawn an embedding diagram representing a slice through the middle of the two-sided black hole of Figure 14.1. This is just like the wormhole diagrams in Chapter 6. The two CTFs live on the circles labelled L and R (these circles are points on the left and right vertical lines in Figure 14.1). RT says that the entanglement entropy between the quantum theory on L and the quantum theory on R is equal to the size of the smallest curve that divides the interior space into two. In other words, if there is no entanglement, the entanglement entropy is zero and there is no dividing curve and no wormhole. In that case, the two quantum theories are disconnected and there is no space linking them. With maximal entanglement, the wormhole appears and

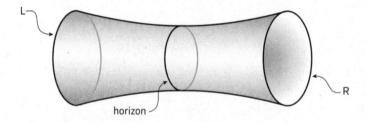

Figure 14.2. A snapshot in time illustrating the wormhole in Figure 14.1. Points on the boundaries, L and R, are circles and the interior is a two-dimensional surface. The horizon is the smallest curve that divides the wormhole in half. The length of the horizon fixes the amount of quantum entanglement between the two quantum theories living on L and R.

the smallest curve is as drawn in Figure 14.2: it is the horizon and it wraps around the wormhole at its narrowest point.

This result should ring bells. If we make things a bit more realistic by adding another dimension of space (we cannot visualise the wormhole now), the CFTs live on spheres that bound a three-dimensional space. The entanglement entropy between the two CFTs is equal to the area of the throat of the wormhole that connects them, and recall from Chapter 6 that, when the wormhole is at its shortest, this is equal to the area of the event horizon of the black hole. This sounds very much like Bekenstein's result: the area of the event horizon of a black hole is equal to its thermal entropy. Here, the RT conjecture is telling us that the area of the event horizon is given by the entanglement entropy between two quantum theories.

This conclusion is worth repeating. We started out with two isolated quantum theories, each describing a bunch of particles. If these theories are not entangled, the two theories describe two disconnected universes. As in the previous chapter, the two theories still each have their own holographic dual but they are otherwise entirely disconnected. If instead we set up the mathematics so that the two theories are entangled with each other, holography tells us that the dual description is a wormhole. And the RT conjecture relates the entanglement entropy between the two quantum theories, which we can calculate using quantum theory alone, to the geometry of the wormhole – and in particular to the area of the wormhole at its narrowest point, which is also the area of the event horizon of the black hole.

Entanglement makes space

Although these links between entropy, entanglement and geometry were initially discovered in the context of black holes, they are now understood to be much more general. In his 2010 prize-winning essay entitled 'Building up spacetime with quantum entanglement', Canadian physicist Mark Van Raamsdonk writes that 'we can connect up spacetimes by entangling degrees of freedom and tear them apart by disentangling. It is fascinating that the intrinsically quantum phenomenon of entanglement appears to be crucial for the emergence of classical spacetime geometry.' By 'degrees of freedom', Van Raamsdonk is referring to particles, qubits or whatever are the 'moving parts' of the quantum theory, and by 'connecting up' or 'tearing apart' spacetime, he means that entanglement is not only related to geometry – it underlies it. Here's the idea.

At the top left of Figure 14.3 we show a sphere, and on the boundary of the sphere is a quantum theory in its vacuum state. The vacuum, if you recall, is highly entangled. We've split the boundary into two parts, labelled L and R. The vacuum on the left part of the boundary is entangled with the vacuum on the right. The RT conjecture says that the amount of entanglement between these two regions on the boundary is equal to the area of the smallest possible surface (known as the 'minimal surface') that

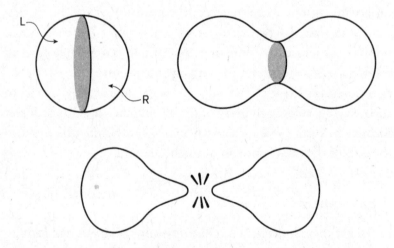

Figure 14.3. As the entanglement between the two halves of the sphere decreases, the bulk space stretches apart and eventually splits into two disconnected regions.

correspondingly divides the interior space. This dividing surface is shown as the shaded disk. There are no black holes here – just space bounded by a sphere. Suppose now that we could reduce the amount of entanglement on the boundary. According to RT, the area of the surface splitting the two regions must also reduce, which means that the two halves are joined together as in the top right picture. If the entanglement is further reduced to zero, the interior splits into two disconnected regions, as illustrated in the bottom picture. Space exists only in the interior regions of each bubble; there is no space connecting the bubbles. Thus, we see that the geometry of the interior space – the bulk – changes as we change the amount of entanglement in the quantum theory on the boundary of the space. But, as Einstein taught us, the geometry of space is gravity. This is the remarkable essence of the RT conjecture: gravity is determined by entanglement.

As an aside, the RT conjecture also delivers insight into something we said in the context of the AMPS firewall paradox. The essence of the paradox concerned the effects of breaking the entanglement of the quantum vacuum across the event horizon. This, we claimed, would be tantamount to tearing open empty space. We see this effect in a different guise in Figure 14.3. Here, if we switch the entanglement off between two regions of the quantum theory on the boundary, we slice the interior space in two.

We may also be glimpsing something even more deeply hidden: a new way of thinking about quantum entanglement. Let's think not about entanglement between two CFTs on the boundary of space, but the much simpler case of entanglement between two particles. The idea, which has become known as the ER = EPR conjecture, asserts that we can picture these two particles as being linked together by something akin to a wormhole. This 2013 conjecture, due to Juan Maldacena and Leonard Susskind,[47] follows nicely from Van Raamsdonk's work. The ER side of the equation refers to the Einstein–Rosen bridge (wormhole) and the EPR side refers to the famous analysis by Einstein, Boris Podolsky and Nathan Rosen in which they attempted to make sense of quantum entanglement.[48] Just as the Einstein–Rosen bridge between two black holes is created by quantum entanglement, so, in Maldacena and Susskind's words, 'it is very tempting to think that *any* EPR correlated system is connected by some sort of ER bridge, although in general the bridge may be a highly quantum object that is yet to be defined. Indeed, we speculate that even the simplest singlet state of two spins is connected by a (very quantum) bridge of this type.'

Islands in the stream*

It's now time to return to black holes and an important thread we have left hanging. Maldacena showed that it is possible for information to emerge from black holes, and that was enough for Stephen Hawking. However, at the time Hawking conceded his bet, nobody knew how the information emerges, and, in a closely related question, nobody knew what was wrong with Hawking's original 1974 calculation. Until 2019, this was how things stood in the theoretical physics research community. The breakthrough came from two independent groups who were able to derive the all-important Page curve using 'old fashioned' physics (general relativity and quantum mechanics).[49] The calculations support the holographic notion that the distant Hawking radiation and the interior Hawking radiation are two versions of the same thing. It is remarkable that the Page curve can be derived using 'old fashioned' physics, and another hint that Einstein's theory of gravity knows much more about the fundamental workings of Nature than we might otherwise have given it credit for. The laws of black hole mechanics that we met in Chapter 10 are another striking

---

* With acknowledgement to T. Hollowood, S. Prem Kumar, A. Legramandi and N. Talwar, of Swansea University, for their 2021 paper (*J. High Energy Phys.* 2021(11):67). And possibly also to Dolly Parton and Kenny Rogers.

example of this hidden depth of general relativity. They reveal to us that the theory knows something about the underlying microphysics, since Hawking's Area Theorem is the Second Law of Thermodynamics in disguise.

The big idea in the 2019 papers is that, for an old black hole (one that is older than the Page time) part of the interior of the black hole is really on the outside. The full ramifications of this inside-outside identification remain to be understood but, as we will now see, both RT and the ER = EPR conjecture play a role.

In Figure 14.4, we show the interesting part of the Penrose diagram of an evaporating black hole.* The Hawking radiation streams along 45-degree lightlike trajectories heading towards future lightlike infinity and the partner particles head along similar trajectories inside the horizon. In Einstein's theory, these partners would be destined to hit the singularity, but in the new calculations something more dramatic happens: the partner particles behind the horizon end up outside.

Let's first see how this leads to the Page curve. Imagine someone far away from the black hole sitting a fixed distance away and collecting the Hawking radiation. This observer follows the wiggly line on the Penrose diagram (you might like to check that by looking at the Schwarzschild coordinate grid on Figure 5.1). Suppose that our observer collects all the radiation emitted up to some time, t. We will refer to this radiation collectively as R. Our interest is in knowing the entanglement entropy between the radiation they collect and the black hole. If t is large enough, the black hole will have evaporated away, and the observer will have collected all the Hawking radiation. In that case, the entanglement entropy should have fallen to zero if all the information came out. This is precisely what happens in the new calculations, and precisely what doesn't happen in Hawking's.

---

* A special thank you to Tim Hollowood for his insight and help, especially concerning this figure and also Figure 14.6.

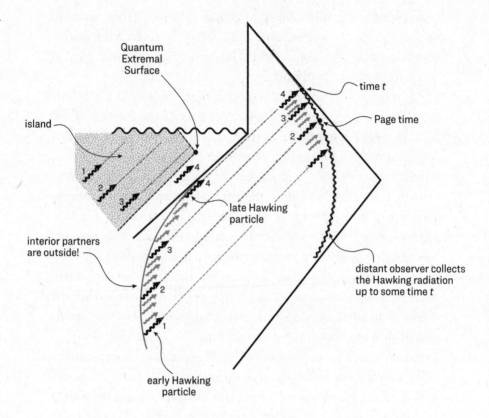

Figure 14.4. Part of the Penrose diagram corresponding to an evaporating black hole. The wiggly arrows denote Hawking particles and particles with the same number are entangled partners (one is outside the event horizon and its partner is inside). The Quantum Extremal Surface corresponding to the radiation R is indicated, along with its island (the shaded region). Interior partners in the island should be considered to be part of R. Also on plate 15.

The essential difference between the two calculations lies in the shaded region inside the horizon marked 'island'. The island is a special region of spacetime. Its existence, and where it is located, is the subject of the 2019 papers. It turns out that the location of the island is dictated by the amount of radiation our observer has collected. If the time t is smaller than the Page time, there is no island. After the Page time, the island appears. How does the island lead to the Page curve?

At the right-hand tip of the island is a point on the Penrose diagram that we've labelled the Quantum Extremal Surface (QES). As with all the Penrose diagrams we've drawn, this point corresponds to a spherical surface in space.* The modern calculations give us a formula to compute the entanglement entropy in terms of the area of this surface:

$$S_R = \frac{\text{Area of QES}}{4} + S_{SC}$$

S_{SC} is the entanglement entropy of the Hawking radiation just as Hawking computed it, with a very important difference. The calculation mandates that we should also include Hawking partner particles inside the island in the calculation. This is the big new idea. In Hawking's original calculation, he missed the existence of the island. For times before the Page time, when we have less than half the radiation, there is no island and Hawking's calculation is correct. This gives the rising part of the Page curve, illustrated again in Figure 14.5. After the Page time, the island appears, with its QES very close to the horizon.

The new idea tells us that when we calculate the entanglement entropy, we must include the Hawking particles inside the island. These particles, inside the black hole, are 'reunited' with their partners outside and, once reunited, their combined contribution to the entanglement entropy is zero.†

---

* To appreciate that a point on our Penrose diagram is a spherical surface at a moment in time you need to remember that each point on a Penrose diagram represents not just one point in space at a moment in time but all points in space that have the same Schwarzschild R at a moment in time, which is a sphere.

† When one member of an entangled pair is inside some region and its partner is outside then there is a contribution to the entanglement entropy of the region. In contrast, if both are inside (or outside) the region then they contribute nothing to the entanglement entropy of the region.

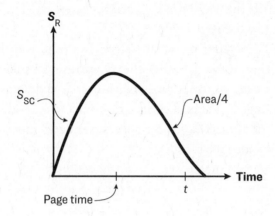

Figure 14.5. The entanglement entropy of the radiation R according to the island formula. Remarkably, it has the same shape as the Page curve (see Figure 12.3).

The result is that, once the island has formed, the overall entanglement entropy of the Hawking radiation, S_R, is given mainly by the first term on the right-hand side of the equation, which is simply the area of the QES (divided by 4). But this is approximately equal to the Bekenstein–Hawking entropy of the black hole because the QES lies close to the horizon. Now, since the area of the horizon shrinks as the hole evaporates, so too does the area of the QES. The entanglement entropy therefore starts to fall, and it goes all the way to zero when the black hole has evaporated away because the area of the QES (and the horizon) vanishes then. In this way, the Page curve starts to fall after the Page time and the calculations deliver the correct Page curve. This is a brilliant piece of physics.

At this point, you may be highly sceptical at what appears to be an egregious sleight of hand. We seem to be reassigning particles inside the event horizon as belonging to the Hawking radiation in order to decrease the entanglement entropy. It is as if we are arbitrarily reassigning pieces of the jigsaw (from Chapter 12) from one table to another. That would be fair criticism if the island formula had been written down merely in order to reproduce the Page curve, but this is not the case. Instead, the island formula can be derived using the same basic physics that Hawking originally employed: quantum physics and general relativity. Hawking simply missed a subtle feature of the mathematics that, once included, triggers the appearance of the island.

One might consider the calculation of the Page curve to be a solution to the black hole information paradox. But we don't intend to leave things at that. We want to know *how* the information gets out. Recent research suggests that the physics is closely related to both the RT conjecture and ER = EPR.

The meaning of the island

There is a striking similarity between the formula for the entanglement entropy of the Hawking radiation and the Ryu–Takayanagi formula. Is there an entanglement-geometry connection here? The answer seems to be yes. In fact, inspired by the ER = EPR conjecture, we might claim that the formula for S_R is precisely the Ryu–Takayanagi result. This is illustrated in Figure 14.6, which also illustrates how it can be that the inside of an old black hole is really the outside.

The top figure shows the situation for a young black hole. It is an embedding diagram, corresponding to the time slice illustrated in the little Penrose diagram on the right.* Notice how the blue and red Hawking partners have been linked by a tiny wormhole: this is the ER = EPR idea. The particles are joined by one of Susskind and Maldacena's highly quantum wormholes. Having made these links, can you see that the distinction between what is outside and what is inside is unclear? The region outside is shaded

---

* The slice is drawn as a slightly wavy line, to indicate it is not unique. What is important is that it never curves upwards at more than 45 degrees. Otherwise, it would not correspond in any sense to 'all of space' at some notion of 'now'.

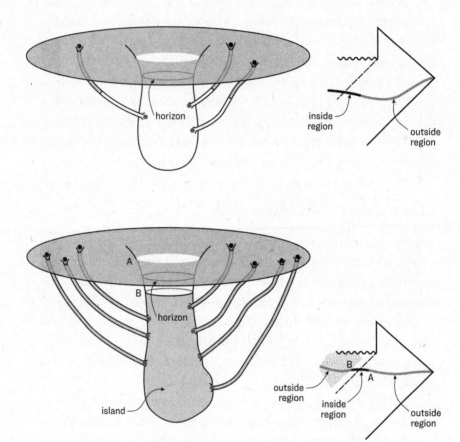

Figure 14.6. Illustrating how the island idea works. It vindicates both the ER = EPR idea and the Ryu–Takayanagi conjecture. For an old black hole (bottom), most of the 'inside' of the black hole is outside. In both pictures, the outside is shaded. Hawking particles are represented by dots (those described as 'red' in the text are black here, and their 'blue' partners are grey). Also on plate 16.

orange and it seems natural to regard all of the flat region at large distances from the hole as being 'outside'. But what about the space inside the wormholes? It seems we can travel from outside to inside along a wormhole. The key question is, where do we draw the line between inside and outside? The answer is given by Ryu–Takayanagi. We should seek the smallest area surface that splits the two regions. For a young black hole, that smallest area is (presumably*) obtained by cutting through the wormholes. None of that is too weird. But for an old black hole there are many more wormholes, and the Ryu–Takayanagi surface is now the QES. This is illustrated by the curve marked B in the lower figure. The region between the QES and the curve marked A is inside, and the remainder, shaded orange, is outside. The island is precisely that part of the interior that should more correctly be regarded as outside.

This is the beginning of a physical picture of how a black hole may return information to the Universe as it evaporates. The picture of the singularity that emerges is more speculative still. As we have sketched the interior of the black hole in Figure 14.6, the singularity appears to be replaced by a quantum network of wormholes connecting to the outside. In Chapter 5, we sent a group of intrepid astronauts into the black hole and they all met their doom at the singularity. If we take this new picture of the black hole at face value, we can ask whether the end of time really does lie in the future of the astronauts. Imagine you are one of the astronauts. You fall across the horizon of the black hole without drama and confront … what? According to Figure 14.6, you will meet a network of wormholes, be dissociated by tidal gravity, scrambled up and the information that is you will emerge, through the wormholes, imprinted in the Hawking radiation.

---

* This interpretation is speculative since we do not understand these little wormholes. The formula for S_R does not rely on this speculation.

15

THE PERFECT CODE

'... every item of the physical world has at bottom – at a very deep bottom, in most instances – an immaterial source and explanation; that what we call reality arises in the last analysis from the posing of yes-no questions and the registering of equipment-evoked responses; in short, that all things physical are information-theoretic in origin and this is a participatory universe.'

John Archibald Wheeler[50]

'... the whole show is wired up together ...'

John Archibald Wheeler[51]

'... time and space are not things, but orders of things ...'

Gottfried Wilhelm Leibniz

Quantum entanglement is turning out to be a key player. We have thought about it so far to keep track of the information coming out of a black hole. In that context, we have seen that entanglement seems to be responsible for creating what we experience as space. What we will learn now is that the way entanglement creates space appears to be very robust. This is just as well for us: we don't want to live in a space that might be prone to falling apart.

Quantum entanglement is also a key resource for those trying to build quantum computers. At first sight, the construction of computing machines might appear to have nothing at all to do with the emergence of space. In a quantum computer, entanglement is the primary means by which information is encoded in a robust way that is resilient to damaging environmental factors. This topic, known as quantum error correction, is fundamental to the construction of working quantum computers. There are parallels here: it is beginning to look as if space is woven out of quantum entanglement in a manner similar to the way quantum engineers weave qubits together to build quantum computers. The suggestion is that there is a link between quantum computing and the fabric of reality. In this chapter we are going to explore that link.

The source code of spacetime

In Figure 15.1, we show a slice through AdS spacetime with an entangled quantum theory on the boundary; the Poincaré disk once more. The boundary has been split into three parts, labelled A, B and C. Let's first focus on region A. Ryu–Takayanagi tells us that the entanglement entropy of region A with regions B and C is given by the length of the shortest line that can divide the two regions. In this AdS spacetime, the shortest line is a curve. Holography tells us that if we know what is happening on the

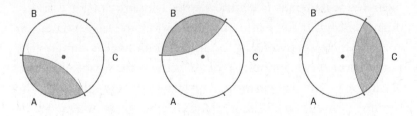

Figure 15.1. Illustrating the causal wedge puzzle.

boundary (A, B and C), we know everything about the interior. It also tells us, and this is not obvious but has been proved, that if we know what is happening on A, we also know what is happening in the shaded region. In the jargon, the shaded region is known as the entanglement wedge of A, because the quantum theory on A entirely determines what happens in the shaded 'wedge'. The same is true for regions B and C, as illustrated in the remainder of the figure. Now consider a point somewhere near to the centre of the disk (the black dot). The left disk tells us that it is encoded on boundary regions B and C. The middle disk says it is encoded on A and C and the right disk says it is encoded on A and B. The only way all three statements can be true is if the information is encoded redundantly. This means that we could erase region A and still know what is happening at the black dot; or we could erase region B or region C. What we can't get away with is erasing two of the three boundary regions – that would be too much. This is intriguing. It means that the answer to the question, 'where on the boundary is the information associated with the region around the black dot encoded?' is 'it isn't in any single region (A, B or C), but the information can be determined from knowledge of any two regions'. It is quantum entanglement that makes this robust distribution of information possible.

According to holography, the information needed to encode the interior space is scrambled up and distributed across the boundary, which makes it hard to read but very robust against destruction. This is very similar to a technique computer scientists have discovered that is central to the construction of working quantum computers. At the time of writing, the largest quantum computers are networks of around 100 entangled qubits. The potential of these computers is vast because the 'space' in which calculations can be performed grows exponentially with the number of qubits, exploiting quantum entanglement as an information resource. These 100-qubit quantum computers can perform calculations in minutes that would take a conventional supercomputer longer than the current age of the Universe to complete.

One of the biggest challenges in building large-scale quantum computers is preventing the qubits from becoming entangled with their environment. Given what we know about entanglement and quantum information, it should be clear that this would be a bad thing because information would 'leak out' of the computer into the surroundings and the computer wouldn't work. Perfect isolation isn't practicable, so what is needed is a way to protect the important qubits that are needed to program the computer: a way to encode information that makes it hard to destroy. This can be done by exploiting quantum entanglement to encode the information in a robust way. This is quantum error correction.

Classical error correction is a routine part of our everyday technology. A QR code, for example, encodes multiple copies of information so that a sizeable part of it can be destroyed while still allowing the information to be decoded. Quantum computers can't rely on storing multiple copies of the information because, as we've seen, the quantum no-cloning theorem prevents quantum information from being copied. The solution is to devise a quantum circuit that encodes the important information in a redundant way without copying, but also in a way that is robust against interactions with the environment. It turns out that the latter is equivalent to requiring that the information should be scrambled up such that, in a sense, it is kept secret from the environment. It is rather like the environment can destroy the precious information only if it understands how we have encoded it. If we scramble things up sufficiently then the environment can't crack the code. We give a non-quantum example of redundant, non-local information encoding in Box 15.1.

BOX 15.1. Encoding information

Suppose we want to encode the three-digit combination to a safe (*abc*). One way to do it is to make use of the function $f(x) = ax^2 + bx + c$. To crack the code, one needs to know the values of *a*, *b* and *c*. It is possible to hide this information among a large group of people by giving each person a pair of numbers; a particular value of *x* and the corresponding value of $f(x)$. To crack the code, we must interrogate any three people in the room for their pairs of numbers *x* and $f(x)$. This is sufficient to determine *a*, *b* and *c*. This secret sharing scheme is a means to redundantly encode information in a non-local way. The method is robust against losing people: so long as we have at least three people we can get the code.

The challenge for those wanting to build a quantum computer is to invent a compact device for encoding a qubit (or a bunch of qubits) inside a bigger block of qubits, so that the qubits we want are safe even if there is damage to the exterior qubits due to their interactions with the environment. Error correction is all about trying to achieve that using an optimal combination of redundancy and secrecy. We can now appreciate the connection with holography, because the coding we've discussed in the context of AdS/CFT is an impressive combination of redundancy and secrecy.*[52] In holography, the boundary codes for the interior space, and it does so in a redundant way because we can erase part of the boundary without losing the information in the interior. It also stores information in a way that is hard to decode, since the information is scrambled up and encoded non-locally by quantum entanglement. To destroy the interior space (as Van Raamsdonk

---

*   The penny dropped first for Ahmed Almheiri, Xi Dong and Daniel Harlow, who pointed out the AdS/CFT link with quantum error correction in 2015.

imagined) we need to destroy the entanglement over a substantial part of the boundary and not just a small part of it.

In 2015, Fernando Pastawski, Beni Yoshida, Daniel Harlow and John Preskill[53] devised an arrangement of networked qubits that redundantly encodes information about the interior of the network on the boundary. This is precisely the situation we've been discussing in the context of holography. The coding is known as the HaPPY code (after the authors' initials) and is shown schematically in Figure 15.2. The open circles around the outside are qubits, as are the circles inside the pentagons. In a quantum computer, the boundary qubits are those most in danger from the environment. The qubits inside the pentagons are the ones the computer will use for its operations, and these are safer because of the structure of the network. The pentagons are devices that entangle the six qubits that feed into them. They operate such that any three qubits are maximally entangled with the other three. This means that the information encoded by the central qubit is robust against the erasure of up to three of the surrounding qubits.

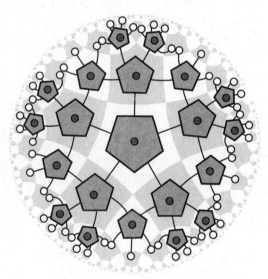

Figure 15.2. The HaPPY holographic pentagon code.

The diagram shows a network with just a few layers of pentagons. You can see (from the underlying shaded pattern) that the pentagons are linked together in a manner that matches the hyperbolic tiling of the Poincaré disk. We could add more layers by moving the external qubits out by another layer. The pentagons would look very small on the diagram but that does not mean they are very small in real life. As physical devices, the pentagons could all be the same size. What matters is the way they are networked together and that is governed by the underlying hyperbolic geometry. This hyperbolic linking is an important feature, as we will now see.

The exciting feature of the HaPPY code is that it reproduces the most important features of AdS/CFT, and in particular the Ryu–Takayanagi result. We illustrate this in Figure 15.3. Each black dot represents a qubit. Let's suppose that we know the state of the 'dangling' qubits around the edge. If lines from three known qubits feed into a pentagon, then we also know the state of the other two qubits, and also the central qubit. Qubits from the outer pentagons link into adjacent interior pentagons. As we head inwards and repeat, we always know the state of all the qubits

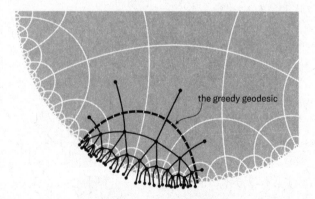

Figure 15.3. The greedy geodesic has a length defined by the number of network legs it cuts through. Starting from the physical qubits dangling on the outside we can move inwards to reconstruct the interior logical qubits shown as black dots inside the pentagons.

linking into each pentagon until we encounter a pentagon with fewer than three inputs. At this point we can't go any deeper; the qubits around the edge no longer encode for that part of the interior. When we reach this stage, the line that we cross is known as the 'greedy geodesic', shown as a dashed line. It marks out the part of the interior that is described perfectly by the dangling qubits around the edge. It is also the shortest line that can be drawn through the interior that links the edges of the boundary region containing the dangling qubits. Remarkably, the amount of entanglement between the qubits in the boundary region and the rest of the boundary is equal to the number of links the greedy geodesic cuts through in the network. This is nothing other than the Ryu–Takayanagi result.

The gem here – the key point – is that the HaPPY code is a network of qubits, and yet it exhibits the properties of the physics we've been discussing in the context of black holes. Try to imagine the HaPPY code without imagining a space in which the qubits are embedded. No space, just entangled qubits. We know that there is an equivalent description of the code using the language of geometry, which is what we've been using to visualise it: it's the hyperbolic geometry of the Poincaré disk. In other words, the way we have wired up the qubits gives rise to an emergent hyperbolic geometry. The notion of distance emerges as the number of links we cut through in the network, i.e. distance is defined by counting the number of links that are cut. Astonishing as it may seem, we are being invited to imagine that the space we live in is built up from an entangled network of elemental entangled quantum units that are too small for us to detect with current experiments. Instead, we are sensitive to the way these entangled units give rise to the physical phenomena we see, including the very idea of space itself. This is quite a remarkable development and one of which John Archibald Wheeler would surely have approved.

So, what is reality?

Are we living inside a giant quantum computer? The evidence is mounting that it may be so. For years, the study of black holes has been an intellectual endeavour that has pushed theoretical physicists into corners. But in the last decade or so, a flurry of understanding, fuelled in large part by exploiting the rapidly developing field of quantum information, has led to a consensus view that holography is here to stay and that it shares many similarities with quantum error correction.

Does living inside a universe that resembles a giant quantum computer suggest that we are virtual creations living inside the computer game of a super-intellect? Probably not. There is no reason to make that link. Rather, in our pursuit of the quantum theory of gravitation, the bluest of blue skies research, we appear to have glimpsed a deeper level of the world, and understanding this deeper level may well be useful to us when we design quantum computers. This has happened so many times in the history of science. We are constantly discovering techniques that Nature has already exploited. It is not so surprising that those techniques turn out to be useful to us technologically: it seems that Nature is the best teacher.

This unlikely link between quantum computing and quantum gravity raises tantalising new possibilities. The future of quantum gravity research may have an experimental side to it, something thought highly unlikely just a few years ago. Maybe we can explore the physics of black holes in the laboratory using quantum computers. And this deep relationship between the two fields flows both ways. There may in turn be a good deal of overlap between pure black hole research and the development of large-scale quantum computers, devices which will be of enormous benefit to our economy and the long-term future of our civilisation. Perhaps it will not be long before we can no more imagine a world without quantum computers than we could imagine a world without classical computers today.

This is the ultimate vindication of research for research's sake: two of the biggest problems in science and technology have turned out to be intimately related. The challenge of building a quantum computer is very similar to the challenge of writing down the correct theory of quantum gravity. This is one reason why it is vital that we continue to support the most esoteric scientific endeavours. Nobody could have predicted such a link.

'Be clearly aware of the stars and the infinity on high. Then life seems almost enchanted after all', wrote Vincent van Gogh. The study of black holes has attracted many of the greatest physicists of the last 100 years because physics is the search for both understanding and enchantment. That the quest to understand the infinities in the sky has led inexorably to the discovery of a holographic universe enchanting in its strangeness and logical beauty serves to underline Van Gogh's insight. Perhaps it is inevitable that human beings will encounter enchantment when they commit to exploring the sublime. But it's bloody useful too.

ACKNOWLEDGEMENTS

We are very grateful for the help and support of numerous colleagues, family and friends. In particular, for their many helpful comments and discussions, we'd like to thank Bob Dickinson, Jack Holguin, Tim Hollowood, Ross Jenkinson, Mark Lancaster, Geraint Lewis, Chris Maudsley, Peter Millington and Geoff Penington. We are also grateful for the support received from the University of Manchester and the Royal Society.

For helping make the book happen, a big thank you is due to Myles Archibald and the team at HarperCollins, and to Diane Banks, Martin Redfern and Sue Rider.

Most of all, we are indebted to our families – to Marieke, Florence, Isabel, Lenny and Tilly, and to Gia, George and Mo.

Thank you all.

ENDNOTES

1. Hawking, S. W. and Ellis, G. F. R. (1973), *The Large Scale Structure of Space-Time* (Cambridge University Press, Cambridge).
2. Einstein, A. (1939), 'On a Stationary System with Spherical Symmetry Consisting of Many Gravitating Masses', *Ann. Math. Second Series*, 40(4):922–936.
3. Montgomery, C., Orchiston, W. and Whittingham, I. (2009), 'Michell, Laplace and the Origin of the Black Hole Concept', *J. Astron. Hist. Herit.*, 12(2):90–96.
4. Fowler, R. H. (1926), 'On Dense Matter', *MNRAS*, 87:114–122.
5. Chandrasekhar, S. (1931), 'The Maximum Mass of Ideal White Dwarfs', *Astrophys. J.*, 74:81–82.
6. Oppenheimer, J. R. and Snyder, H. (1939), 'On Continued Gravitational Contraction', *Phys. Rev. Lett.*, 56:455.
7. Wheeler, J. A. with Ford, K. (2000), *Geons, Black Holes, and Quantum Foam. A Life in Physics* (W. W. Norton & Co., New York).
8. Fuller, R. W. and Wheeler, J. A. (1962), 'Causality and Multiply Connected Space-Time', *Phys. Rev.*, 128:919–929.
9. Penrose, R. (1965), 'Gravitational Collapse and Space-Time Singularities', *Phys. Rev. Lett.*, 14:57.
10. Hawking, S. W. (1974), 'Black Hole Explosions?', *Nature*, 248: 30–31.
11. Wigner, Eugene P. (1960), 'The Unreasonable Effectiveness of Mathematics in the Natural Sciences', *Comm. Pure Appl. Math.*, 13:1, 1–14.

12. Taylor, E. F., Wheeler, J. A. and Bertschinger, E. W. (2000), *Exploring Black Holes* (Pearson, New York).

13. Page, D. N. (2005), 'Hawking Radiation and Black Hole Thermodynamics', *New J. Phys.*, 7:203.

14. Hawking, S. W. (1975), 'Particle Creation by Black Holes', *Comm. Math. Phys.*, 43:199–220.

15. Misner, C. W., Thorne, K. S. and Wheeler, J. A. (1973), *Gravitation* (Princeton University Press, Princeton).

16. Hafele, J. C. and Keating, R. E. (1972), *Science*, 177(4044):168.

17. Misner, C. W., Thorne, K. S. and Wheeler, J. A. (1973), *Gravitation* (Princeton University Press, Princeton).

18. Hamilton, A. J. S. and Lisle, J. P. (2008), 'The River Model of Black Holes', *Am. J. Phys.*, 76:519–532.

19. Einstein, A. and Rosen, N. (1935), 'The Particle Problem in the General Theory of Relativity', *Phys. Rev.*, 48:73.

20. Taylor, E. F., Wheeler, J. A. and Bertschinger, E. W. (2000), *Exploring Black Holes* (Pearson, New York).

21. Morris, M., Thorne, K. and Yurtsever, U. (1988), 'Wormholes, Time Machines and the Weak Energy Condition', *Phys. Rev. Lett.*, 61(13):1446–1449.

22. Hawking, S., Thorne, K., Novikov, I., Ferris, T., Lightman, A. and Price, R. (2002), *The Future of Spacetime* (W. W. Norton & Co., New York).

23. Droz, S., Israel, W. and Morsink, S. M. (1996), 'Black Holes: the Inside Story', *Phys. World*, 9(1):34.

24. Chandrasekhar, S. (1987), *Truth and Beauty* (University of Chicago Press, Chicago).

25. Wheeler, J. A. with Ford, K. (2000), *Geons, Black Holes, and Quantum Foam. A Life in Physics* (W. W. Norton & Co., New York).

26. Abbott, J. (1879), 'The New Theory of Heat', *Harper's New Monthly Magazine*, XXXIX.

27. Atkins, P. (2010), *The Laws of Thermodynamics: A Very Short Introduction* (Oxford University Press, Oxford).

28. Letter to John William Strutt, Baron Rayleigh, dated 6 December 1870.

29. Goodstein, D. L. (2002), *States of Matter* (Dover Publications, New York).

30. Feynman, R. P. (1997), *The Character of Physical Law* (Random House, New York).

31. Hawking, S. W. (1974) 'Black Hole Explosions?', *Nature*, 248: 30–31.

32. Bardeen, J. M., Carter, B. and Hawking, S. W. (1973), 'The Four Laws of Black Hole Mechanics', *Comm. Math. Phys.*, 31(2):161–170.

33. Hawking, S. W. (1974) 'Black Hole Explosions?', *Nature*, 248: 30–31.

34. From the Proceedings of the third International Symposium on the Foundations of Quantum Mechanics, Tokyo, 1989.

35. Fulling, S. A. (1973), 'Nonuniqueness of Canonical Field Quantization in Riemannian Space-Time', *Phys. Rev., D.*, 7(10): 2850.
Davies, P. C. W. (1975), 'Scalar Production in Schwarzschild and Rindler Metrics', *Phys. A.*, 8(4):609.
Unruh, W. G. (1976), 'Notes on Black-hole Evaporation', *Phys. Rev. D.*, 14(4):870.

36. Black hole complementarity was first introduced in a paper written in 1993 by Leonard Susskind, Lárus Thorlacius and John Uglum, following earlier work by Gerardus 't Hooft.
Susskind, L., Thorlacius, L., Uglum, J. (1993), 'The Stretched Horizon and Black Hole Complementarity', *Phys. Rev. D.*, 48(8):3743.
't Hooft, G. (1990), 'The Black Hole Interpretation of String Theory', *Nucl. Phys. B.*, 335(1):138.

37. Susskind, L. (2008), *The Black Hole War* (Little Brown, New York).

38. Nielsen, M. A. and Chuang, I. L. (2010), *Quantum Computation and Quantum Information*, (Cambridge University Press, Cambridge).

39. Kwiat, P. G. and Hardy, L. (2000), 'The Mystery of the Quantum Cakes', *Am. J. Phys.*, 68(1):33–36.

40. Page, D. N. (1993), 'Information in Black Hole Radiation', *Phys. Rev. Lett.*, 71:3743.

41. Almheiri, A., Marolf, D., Polchinski, J. and Sully, J. (2013), 'Black Holes: Complementarity or Firewalls?', *J. High Energy Phys.*, 2013(2).

42. Hayden, P. and Preskill, J. (2007), 'Black Holes as Mirrors: Quantum Information in Random Subsystems', *J. High Energy Phys.*, 2007 (9):120.

43. Susskind, L. (1995), 'The World as a Hologram', *J. Math. Phys.*, 36:6377–6396.

44. Maldacena, J. (1998), 'The Large N Limit of Superconformal Field Theories and Supergravity', *Adv. Theor. Math. Phys.*, 2(4): 231–252.

45. Penington, G. (2020), 'Entanglement Wedge Reconstruction and the Information Paradox', *J. High Energy. Phys.*, 2020(9):2.

46. Ryu, S. and Takayanagi, T. (2006), 'Aspects of Holographic Entanglement Entropy', *J. High Energy Phys.*, 2006(8):045.

47. Maldacena, J. and Susskind, L. (2013), 'Cool Horizons for Black Holes', *Fortsch. Phys.*, 61:781.

48. Einstein, A., Podolsky, B. and Rosen, N. (1935), 'Can Quantum-mechanical Description of Physical Reality be Considered Complete?', *Phys. Rev.*, 47(10):777.

49. Almheiri, A., Engelhardt, N., Marolf, D. and Maxfield, H. (2019), 'The Entropy of Bulk Quantum Fields and the Entanglement Wedge of an Evaporating Black Hole', *J. High Energy Phys.*, 2019(12):63. Penington, G., 'Entanglement Wedge Reconstruction and the Information Paradox', *J. High Energy Phys.*, 2020(9):2.

50. From the Proceedings of the third International Symposium on the Foundations of Quantum Mechanics, Tokyo, 1989.

51. ibid.

52. Almheiri, A., Dong, X. and Harlow, D. (2015), 'Bulk Locality and Quantum Error Correction in AdS/CFT', *JHEP*, 04:163.

53. Pastawski, F., Yoshida, B., Harlow, D. and Preskill, J. (2015), 'Holographic Quantum Error-correcting Codes: Toy Models for the Bulk/Boundary Correspondence', *JHEP*, 06:149.

PICTURE CREDITS

All Penrose diagrams by Martin Brown and Jack Jewell and all other illustrations by Martin Brown and © HarperCollins*Publishers*, with the following exceptions:

Figure	Credit
1.1	European Southern Observatory/EHT Collaboration/Science Photo Library
3.1	M.C. Escher's *Hand with Reflecting Sphere* © 2022 The M.C. Escher Company-The Netherlands. All rights reserved. www.mcescher.com
4.1	meunierd/Shutterstock
5.4	Wendy lucid2711/Shutterstock, annotated by Martin Brown
5.6	Film still from *Father Ted*, series 2, episode 1 © Hat Trick Productions
6.8	Illustrations by Jack Jewell
7.8	Figure 33.2 from *Gravitation* by Charles W. Misner, Kip S. Thorne and John Archibald Wheeler, page 908. Published by Princeton University Press in 2017. Reproduced here by permission of the publisher.
8.1	Reprinted figure with permission from as follows: Roger Penrose, 'Gravitational Collapse and Space-Time Singularities', *Physical Review Letters*, vol. 14, iss. 3, page 57, 1965. Copyright 1965 American Physical Society

INDEX